War, Virtual War and Society

The Challenge to Communities

War, Virtual War and Society

The Challenge to Communities

Edited by

Andrew R. Wilson
and Mark L. Perry

Amsterdam - New York, NY 2008

The paper on which this book is printed meets the requirements of "ISO 9706:1994, Information and documentation - Paper for documents - Requirements for permanence".

ISBN: 978-90-420-2347-5
©Editions Rodopi B.V., Amsterdam - New York, NY 2008
Printed in the Netherlands

Contents

Part IV Parallels

Notes on Contributors   169

Welcome to an *At the Interface/Probing the Boundaries*
Publication

By sharing insights and perspectives that are both inter-disciplinary and multi-disciplinary, *ATI/PTB* publications are designed to be both exploratory examinations of particular areas and issues, and rigorous inquiries into specific subjects. Books published in the series are enabling resources which will encourage sustained and creative dialogue, and become the future resource for further inquiries and research.

War, Virtual War and Society emerges from the *War, Virtual War and Human Security* project. The aim of the project is to provide an innovative, cutting-edge inter-disciplinary research forum which will enable multiple insights and perspectives to be brought to bear on the many and various issues which relate to war and virtual war. In particular, the project will focus on the nature, purpose and experience of war, and its impacts on all aspects of communities across the world. Viewing war as a multi-layered phenomenon, the project will explore the historical, legal, social, religious, economic, and political contexts of conflicts, and assess the place of art, journalism, literature, music, the media and the internet in representation and interpretation of the experience of warfare. Key themes will include

- the sources, origins, and causes of war
- the 'control', conduct and limit of warfare
- the nature of warfare; types of warfare
- strategy, strategic thinking and the influence and effect of technologies
- war crimes and crimes against humanity
- the extent of war; blockades, sanctions, defence expenditure and the impact on social and public policy
- the 'ethics' of war; just war, deterrence, defence and self-defence, collateral damage
- the experience of war; art, literature, music and the theatre
- the role of the media - journalism, radio, television, the internet
- the prevention of war; the role of conflict resolution, peace-keeping and the role and importance of law
- the rise and impact of non-violent movement

Dr. Robert Fisher
Inter-Disciplinary.Net
www.inter-disciplinary.net

Introduction

In October of 2004 a disparate group comprised of representatives from academe, government, the military and business gathered in Salzburg, Austria, for an intensive three-day conference on the relationship between war and society. This volume represents much of the content of that conference and the discussions that followed.

It is axiomatic to say that war transforms society and militaries are shaped by the societies from whence they are recruited. In recent decades, a significant portion of military studies has in fact been devoted to exploring the technological aspects of this linkage: how peacetime innovations transform the conduct of war and how wartime developments have peaceful applications. This dichotomous approach, however, neglects other dialectical linkages between the battlefront and the home front. While conflict may take place at far remove from the town square, war and society can never be completely isolated from each other. This assertion is more persuasive in our own age when with unprecedented ease war comes into our living rooms or onto our computer screens. But by no means is this solely a product of modern information technology. Throughout history the back and forth linkages between battlefront and home front have been robust and mutually transforming.

The essays in this volume explore this dialectical theme in history and in contemporary conflicts. They also aim to broaden the discussion of historical memory; victor, vanquished and victim; and crime and competition in the context of war and virtual war. The first group of papers looks at the experiences of the home front during the Great War. The second group deals with the memories and ghosts of war. In the third section, we explore information technology not merely as an influence on the conduct of war but also as a force that transforms the very nature of war itself. In the final section, we turn to the social roles that are created by war and to the parallels between the roles and constructs of war and business.

War in real time is an intensely personal experience, but as historical memory it often becomes a collective experience which subsumes individual narratives. The collective experience of World War I is best demonstrated by Paul Fussell's classic *The Great War and Modern Memory* and more recently by Niall Ferguson's *The Pity of War*. In all three of our essays on that conflict, however, we see the varied experiences of the war as it occurred. Both Flothow and Horn reveal the self-reflexive quality of popular culture during the war and the immediacy of cultural production about the war. War as represented in Flothow's children's novels and Horn's cartoons is at once bizarre and intimately familiar. Riez examines how national mobilization and the immense war effort impacted the social, economic and cultural fabric of Hungary's capital, Budapest

Complicating the manifold experiences of war are the paired issues of grief and reconciliation in war's aftermath. Our second section opens with Le Juez's examination of the film *Hiroshima mon amour*. Set against the backdrop of Hiroshima, a French woman - cast out of her hometown for a liaison with a German soldier - and a Japanese veteran come to grips with their grief by sharing individual stories of loss and suffering. In the process they uncover the identities that their personal and collective traumas had obscured. Maillot's chronicle of the difficult, if not impossible, task of reconciliation in contemporary Northern Ireland highlights the additional challenges of terminating a domestic conflict in which the battle lines are blurred and the distinctions between combatants and non-combatants, and between victims and perpetrators, are not only cloudy but contested.

It is always unclear how a new military instrument will influence the conduct of war. Such was the case with strategic bombing in the period between the World Wars and the same is true with information technology today. What we can see is a familiar pattern of doctrinal and strategic experimentation prior to a war, further shaped by interaction in war and reinforced by the proliferation of new technologies. Jokisipilä's examination of cyberwar in the Middle East introduces new players in state-on-state and state-non-state conflict. Whereas the anarchists of the early twentieth century used dynamite and the press, today's cyber-warriors find it easy to escalate a conflict, at least horizontally, via the Internet. It remains to be seen, however, if this is truly a new instrument of war - that is, one able to cause physical, emotional, and economic damage on a scale comparable to more traditional means - or if it will remain a tool primarily of political mobilization and propaganda. Perry, on the other hand, examines how information technology enables society's grass-roots level to monitor conflict and constrain the actions of belligerents through the power of a collective "gaze."

Equating war and business is very much in vogue, as the surfeit of pop-strategy guides available in print and on the web will attest. Seeing parallels between peacetime occupations and military imperatives is, however, nothing new. With the paired transformations of warfare and society that followed the French Revolution, it was natural to see linkages between the routinisation and rationalization of commerce and the requirements of fielding and supplying modern armies. As we see in the contribution of Fritzsche, this trend only expanded with the introduction of organizational and systems theories following World War II. Meyer and

Wilson describe a similar trend at work in ancient China. The author of *Sun Tzu's Art of War* (the classic of strategy that is most frequently applied to the world of business) advocated new social and institutional roles for a "professional" general that mirrored the bureaucratic rationalization making possible the mass warfare of fourth century BCE.

Rarely do academics and policymakers have the opportunity to sit down together and contemplate the broadest consequences of war. Our comprehension has traditionally been limited to war's causes, execution, promotion, opposition, and immediate political and economic ends and aftermath. But just as public health researchers are becoming aware of unexpected, subtle and powerful consequences of human economic action, we are beginning to realize that war has many short- and long-term consequences that we poorly understand but cannot afford to neglect. War influences attitudes among children, conceptions of popular culture and socializing habits vital to the cosmopolitan ideal. It causes chronic long-term social disunity, psychological distortions, re-conceptions of the information revolution and globalization, and fundamental changes in the experience of social reality and in the practice of leadership.

These papers contribute to a growing discourse among academics, scholars and lawmakers that is questioning and rethinking the nature and purpose of war. By studying the effects of war on communities we can more readily understand and anticipate the consequences of present and future conflicts. Such an understanding might well enable us to plan and execute military action with a more clearly defined set of post-war goals in mind. Whereas traditionally a government at war seeks the defeat of the adversary as its primary and often sole aim, through a clearer understanding of war's effects other aims will also become prominent. War, like surgery, could gradually become more refined, could minimize damage in ways that are currently unimaginable, and could involve an increasingly heavy responsibility to prepare for and facilitate reconstruction.

Working counter to this positive trend, the proliferation of new tactics and instruments - be they weapons of mass destruction, information technology or the rudimentary improvised explosive devices that have so plagued coalition forces in Iraq and Afghanistan - has given individuals and sub-state actors the ability to initiate and escalate wars outside the state system. As these new weapons and tactics are paired with radical ideologies and ethno-religious passions, the challenges and dangers facing communities and governments may only increase. War, then, can become more anarchic and less discriminate in its targeting. The sword of transformation cuts both ways and we must anticipate both contingencies.

Projects such as this volume are, of course, only the beginning. Like everything else in this age of globalisation, war is being transformed. The butterfly effect described in chaos theory is amply demonstrated in the radical waves of change passing through and between the military and civilian worlds. The tool that we had held in our hands for so long is now unfamiliar; and in order to achieve our political and civic goals we are obliged to reacquaint ourselves constantly with it, since it may well continue to change into the foreseeable future. The greatest students of war, among them Sun Tzu, Clausewitz, Machiavelli and Mao Tse-tung, were those that recognized shifts in the existing military paradigm as they were happening. To understand this process in the past hones our ability to recognize such shifts in our own age. In addition, given the inherently violent nature of war, the search for such insight is a moral imperative for scholars, diplomats and military commanders alike. The more we understand the evolving nature of war, the better prepared we will be to protect communities from its harmful effects.

Part I

World War I

"Train Yourselves to Defend Your Country": British Children's Novels in the First World War

Dorothea Flothow

Abstract: This paper provides an analysis of children's war novels published during the years of the Great War. It deals with the question of how this war, which, due to its scale and length, affected British society to a much larger extent than the wars of the previous decades, was presented and explained to a youthful readership. It will be argued that children's war novels were a means of propaganda to convince readers of the righteousness and justice of the British cause. Novelists depicted the war as an exciting and basically harmless affair. Enemies, especially the Germans, were presented as the embodiment of evil. Novelists emphasized the glorious and heroic side of war, downplaying its terrible and cruel aspects. In order to do so, they not only took up themes from contemporary wartime propaganda, but also made use of the narrative strategies, motifs and metaphors they had been using to describe previous conflicts, thereby attempting to ignore what was perceived as the specific and "modern" characteristics of the Great War.

Key Words: Children's War Novels, World War I, War Propaganda, Enemy, Narrative Strategies, Myth of Disillusionment

1. Introduction

> Frigidly she bowed to him, took her hand from her muff, and with another curt bow presented him with a white feather.
> "Just to remind you," said she, "that your place is in the trenches and not lolling and flirting over the counter of a London novelty shop."
> With another sarcastic bend of her head she walked out, leaving Harold dumb with anger and amazement, and the white feather, the badge of cowardice, in his hand.[1]

Of course Harold, the hero of Tom Bevan's children's novel *Doing His "Bit"* (1917), in no way deserves the white feather he has been given by the stranger. For when he heard of the outbreak of the First World War, he immediately returned to Britain in order to enlist (which he was about to do when he received the feather). In his eagerness to do his "bit" he resembles all of the other heroes and heroines of the innumerable children's war novels published in the years of the Great War. These stories tell of the exciting adventures and great deeds of a boy hero or a girl heroine against the backdrop of this conflict. As in those war novels for children written in the decades preceding the Great War, the heroes and heroines usually take part in some of the most significant events and battles of the conflict.[2]

This paper is based on an analysis of about 50 of these novels.[3] They were written for both boys and girls and published in book form as well as in magazines, thus reaching a very large audience. The aim of these novels was not merely to entertain their readers: they were also an attempt to explain to them what the war was about, why Britain was fighting it and, ultimately, to gain their support. Quotations like the following, which is taken from the preface to Escott Lynn's novel *In Khaki for the King* (1915), are proof of this aim:

> It is up to you boys of to-day to see that a similar danger
> never threatens your glorious Empire again. Those of
> you who have not yet donned khaki, see to it that when
> you are old enough you train yourselves to defend your
> homes, your mothers, your sisters, your country, all that
> you hold dear.[4]

The question, then, is how the novelists under discussion attempted to achieve these aims. Two basic strategies can be distinguished to this end: on the one hand, novelists took up contemporary wartime propaganda; on the other they also made use of narrative strategies, plot structures and metaphors that they had already been using in pre-world war novels. As has been pointed out by various studies, many late nineteenth- and early twentieth-century children's novels advocated imperial expansion and displayed strongly militaristic attitudes.[5] They celebrated heroism and soldierly virtues and painted highly biased and idealized pictures of past wars. In this, they mirrored society's predominant attitude towards war, for although there were of course groups who were opposed to militarist attitudes - for instance parts of the working classes - militarism was a widespread phenomenon which was not solely confined to the cultural and political elite.[6] In order to depict war as exciting but harmless, writers of children's war novels developed a number of stereotypical methods and strategies, by which the dark side of war was played down or suppressed. These methods and conventions were now adopted to the same end again, the result being that even the Great

War was presented as an exciting adventure, not as the terrible mass slaughter which it undoubtedly was. Moreover, the authors tried to create the impression that the First World War was really not very different from the small heroic conflicts they had depicted in the novels of the pre-world war era. Thus, novelists attempted to counteract the negative image of the War that children might receive through relatives at the front or from their own experience in it.

2.   Novels as Propaganda

The reason why children's novelists were so eager to justify the Great War to their audience seems obvious: The First World War was the first "modern" and "total" war in which Britain participated.[7] It was quite unlike the small colonial wars of the past century, in which only a small professional army had been involved and in which casualties on the British side had usually been limited. Now, the support of the population as a whole was necessary, for Britain needed not only a large army but also munitions workers, nurses, mechanics and people to work the land.[8] With the outbreak of war a huge propaganda machine was gradually set up, which involved government agencies, national newspapers and other institutions.[9] British novelists and their writings played an important part in this propaganda machine. With their tales of German atrocities and cheerful stories of life at the front they presented the official propaganda in an easily accessible and seemingly innocent way; well-known examples include Ian Hay's *The First Hundred Thousand* (1915) and Rudyard Kipling's "Mary Postgate" (1915).[10] Thus as early as 1914 the support of a number of famous writers was secretly secured.[11]

Children were primary targets of wartime propaganda, for they too became involved in the war.[12] Some of them took on an active part: the Scouts, for instance, handled important patrol work, while girls knitted socks and many boys volunteered for the army, some of them lying about their age in order to enlist. But children also became passive victims of the war: they suffered from food shortages and air raids and many had to cope with the absence or death of fathers, brothers and friends. Like adults, children were subjected to extensive patriotic propaganda, which was passed on in school lessons and church sermons, as well as through toys and stories about heroic deeds at the front.[13] A similar kind of propaganda was directed at children in the other warring nations, where the need to gain the support of the population as a whole was just as pressing.

Children's novelists did their "bit" to gain the support of their readers and depicted heroes and heroines eager to do what was expected of them. The tasks they undertake are numerous; most of the boy heroes join the army or navy and take part in the most exciting battles of the war - for instance Ypres and Gallipoli.[14] In some cases, girls also join the fighting, though rather against their will, as in the case of Pickles, the main character of Ethel Kenyon's *Pickles; A Red Cross Heroine* (1916), who gets involved in an aerial combat and succeeds in killing a famous German flying ace.[15] Other heroines carry out more conventional tasks - munitions-making, nursing and knitting - but even these are exciting and rewarding. However small their contribution, they are all shown to help the war effort, as Mary, the heroine of Brenda Girvin's *Munition Mary* (1918), emphasizes: "One more hand! What use was one more worker where so many must be wanted? But every *one* mattered. It was all the ones, and ones, and ones that made up the great total."[16]

The claim that the heroes and heroines are doing the right thing in helping their country also becomes obvious from their experiences in the war itself. None of the heroes or heroines experiences a truly terrible war. Instead, they meet with exciting, glorious adventures and funny incidents, and they get to know famous war heroes and form close friendships with their comrades. These aspects are more important than the terrible events, the killing and dying, which also take place. At the end of the novels the heroes and heroines are usually rewarded for their brave deeds. An obvious example of this comes from Escott Lynn's *Oliver Hastings, V.C.* (1916).[17] After a series of exciting adventures, the eponymous hero is awarded the Victoria Cross, his friend Vivian the D.S.O., their servant is presented with the D.C.O., and Oliver's father with the Maltese Cross of the Order of the Bath. In Frederick Brereton's *Under Haig in Flanders* (1917) the hero, who has run away from home and joined the army as an ordinary soldier, becomes a captain and is awarded the Military Cross.[18] Moreover, he marries a squire's daughter. The heroines' rewards are somewhat different, yet they, too, profit from the war. In Dorothea Moore's *Wanted, an English Girl* (1916) Gillian obtains a position at the court of the fictional Grand-duchess of Insterberg, whereas the heroine of Bessie Marchant's *A Girl Munition Worker* (1916) wins the heart of the man she loves.[19] These are just some of many examples by which the novels suggest that this war is not terrible, but rather a glorious and rewarding experience to those Britons who take part in it.

3.   Meeting the German Enemy

Especially when fighting abroad, some of the characters have close contact with the German enemy.[20] Because many of the heroes happen to be in Belgium at the outbreak of the war, they witness the German invasion and the many atrocities committed against the civilian population.[21] The plunder of "poor little Belgium," which had been invaded by the Germans despite the fact that its neutral status had been guaranteed, was one of the most important themes in British propaganda, and was very prominent in children's novels.[22]

An interesting example of this can be found in Escott Lynn's *In Khaki for the King*, which tells the story of Vivian and Oliver. At the outbreak of the war, these two are in Germany, where they discover the German army's plans of attack - an excellent proof, incidentally, that it was the Germans who started the war after years of careful plotting.[23] This accusation was a frequent refrain and implied that Germany alone was responsible for the war, whereas Britain and the other countries were totally surprised by its outbreak. Oliver and Vivian manage to warn the Belgian officials before the Germans arrive and they participate in the first battles of the war. They also witness some of the behaviour of the Germans towards the Belgians. The following quotation describes a massacre of the civilian population, the emphasis being on the deaths of women and children:

> Meanwhile the soldiers made ready to fire, when, just as they were going to pull the trigger, the woman, with a wild cry, rushed up to one of the prisoners, and threw her arms round him as though she would protect him from the bullets...
>
> "Waste no more time; shoot her with the other cattle," Dous cried out; on which the half-drunken soldiers raised their rifles and sent in several volleys.
>
> Bayonets and rifle-butts finished the fearful business, but neither Oliver nor Vivian dared gaze upon the scene. When they again looked out all the victims were dead, even the poor little boy, who had been bayoneted.[24]

This rather crude example is just one of many similar descriptions in which the novels tell of dead civilians, devastated buildings and the rowdy behaviour of the German soldiers occupying Belgium. Similarly, in *Wanted, an English Girl* the heroine discovers a dead baby and mother who have been murdered by the Germans.[25]

The Germans, it is quite evident, are described as cruel, brutal and barbarous "Huns." They represent "brute force" and are referred to as "the enemy of mankind and civilisation" and the "inhuman foe," which the British must defeat in order to bring peace to the world. In accordance with contemporary propaganda, the Kaiser, above all, is made personally responsible for the most terrible deeds committed by the Germans. In

Robert Leighton's *Dreadnoughts of the Dogger* (1916) the Kaiser orders the sinking of a large ocean liner - an incident similar to the sinking of the *Lusitania*, an event which was given great prominence in wartime propaganda. [26]

While the Germans are presented as evil, militaristic "tyrants," British soldiers are the embodiment of heroic and soldierly virtues. Germany represents "Teutonic militarism" at its worst and is accused of starting the war in order to rule the world.[27] Britain, on the other hand, "was working tooth and nail for peace" and joined the war for the sole purpose of protecting weak and oppressed nations.[28] This polarisation serves the obvious purpose of presenting the Great War as yet another battle in the eternal struggle between good and evil.[29] It is used to justify the British cause and to put an acceptable face on the inevitable loss of life. The victims, then, are interpreted as necessary if good is to prevail.

The polarisation of good and evil, which can frequently be found in the propaganda of many different wars, also serves the purpose of making it easier to kill the enemy, who is presented not only as evil but as dehumanised.[30] Consequently, the Germans are regularly described as mad fanatics (for instance Hilde von Einem in John Buchan's *Greenmantle*) or as ugly, unscrupulous criminals.[31] Often, they are merely perceived as "a grey mob," "grey swarms" or a "human avalanche," and not as individual human beings.[32] Their deaths offer little cause for regret; rather, it is suggested that they deserve their fate.

4.   Using Familiar Metaphors and Methods of Describing War

Perhaps the most serious reproach made against the Germans - and this is another theme taken up from British wartime propaganda - is that their way of fighting is deeply unfair. They wage war against women and children, they use poison gas, and they sink merchant ships.[33] In this they differ greatly from the British, whose methods of warfare are depicted as impeccable. British soldiers carefully protect women and children, they never attack an unsuspecting or unarmed enemy, and British gas, unlike German gas, does not kill but merely stupefies.[34]

Whereas the British are careful to follow the rules of "civilised warfare,"[35] the Germans claim that there are no such rules. The following is taken from Bevan's *Doing his "Bit"*:

> Herr Lieutenant, there is a proverb in English - "All's
> fair in love and war." You *say* that, we *believe* it; that is
> the difference between you and us....War is a big game
> - the biggest of all games; and one side must win and
> the other lose. He wins who uses the most and strongest

> means to an end; to win you must destroy... That is
> where you English do not understand what war is. You
> would play it according to rules like your cricket and
> football; you would put a belt around a nation as you do
> around your boxers and say you must not hit "below the
> belt"; you would pad the gloves. Himmel! what folly,
> what a nation of fools![36]

The German officer's speech clearly shows that he and his countrymen have no notion of "playing the game," unlike the British.

The ideal of "playing the game" is of course one which was already well known from pre-war literature.[37] It has its origin in the so-called "games ethic" as it developed in the English public schools and found its most famous expression in Henry Newbolt's poem "Vitai Lampada" with its lines "Play up! Play up! And play the game!"[38] Playing the game, that is the "unquestioning adherence to the rules," became a term frequently used to describe correct and acceptable behaviour both on the playing field and the battlefield.[39] Like other expressions denoting "sporting" behaviour (for instance "to be a sport" or "it's not cricket"), "playing the game" was common in pre-world war literature and was often used in the children's war novels of those decades.[40] Novelists writing in the years of the Great War continued to use the expression - which can be found in May Wynne's *An English Girl in Serbia* (1916 - "'Arrah, the dirst bastes,' cried O'Mara in disgust, 'don't they iver have the laste notion av playin' the game? Shame on them, vid their four to one against a broth av a boy.'"[41]

There are other ways in which pre-world war literature linked war to sports. Stevenson's *Treasure Island* (1883) describes the doctor's reaction to the bombardment by the pirates:

> All through the evening they kept thundering away. Ball
> after ball flew over or fell short, or kicked up the sand in
> the enclosure;... and though one popped in through the
> roof of the long-house and out again through the floor,
> we soon got used to that sort of horse-play, and minded
> it no more than cricket.[42]

The implications of linking war with sports are obvious: by comparing the unknown to the known, war loses much of its strangeness. War becomes part of the reader's everyday world, something trivial that he or she should not fear. Moreover, the novels imply that just like sports, war is regulated by rules. It is therefore quite harmless. The plot of novels such as G.A.

Henty's *The Dash for Khartoum* (1891) shows this clearly: it depicts a group of boys first as members of a school football team and later in the army. In both activities, the better team invariably wins.[43] Of course, the better team always happens to be British.

Novelists describing the Great War continued to use the war-sports metaphor, which Robert MacDonald refers to as "the ultimate euphemism."[44] It was only one of several motifs and narrative techniques which they employed in the years before the war, and continued to employ, in order to cut war down to size. For instance, a motif often used in the pre-war novels was that of disguise. It was a means of creating excitement and of giving the hero opportunities for adventure. In Charles Gilson's *The Lost Column* (1909) the hero and his friends are caught up in the Boxer Rebellion. Gerald, who speaks fluent Chinese, disguises himself as a Boxer in order to reach the European forces.[45] Unfortunately, he is recognized by one of the enemies. In the days following, Gerald experiences an exciting chase through enemy country. This resembles a game of hide-and-seek in which Gerald always manages to outwit his opponent. The same motif, which again trivializes war by linking it to children's play, is frequently used in novels of the First World War. It occurs, for instance, in *Greenmantle* (1916), where Richard Hannay travels to Germany disguised as a Boer in order to discover Greenmantle's identity.[46] He is found out, and in an exciting flight, during which he changes his disguises several times, succeeds in crossing the frontier to safety. His escape is the result of the resourcefulness and cleverness of the typical British hero and is clear proof of British superiority.

These were not the only narrative conventions and images which novelists retained from pre-world war children's novels in order to trivialize war. In novels such as *The Dash for Khartoum* and H.A. Vachell's *The Hill* (1905) war was frequently described as a test which brought out the best in people and turned boys into men.[47] This kind of thinking, which was dominant in the late nineteenth and early twentieth centuries, had its origin in Social Darwinism.[48] According to that theory, war "functioned like well-managed pruning shears, eliminating the weak and undesirable shoots and allowing the development of luxuriant blooms."[49] As true Britons, fictional heroes rise to the challenges of war with astonishing ease and great self-assurance, a fact which also holds true for the heroes of wartime novels. In *Under Haig in Flanders*, Roger's newly acquired manhood, which is the direct result of his training for war, is obvious not only to his friend Bill but to all who look at him:

> It's made a real man of you. Why, you've come on
> wonderful. You've put three inches on round the chest,

your shoulders have spread so that you've had to have a
new issue of uniform, and bless if you ain't got
heavier... You've just muscle and brawn, and are in
first-rate trim for fighting.[50]

The heroes' many medals and decorations are further proof of their
success in meeting the challenge of the First World War. Despite the fact
that the Great War differed in both quality and scale from earlier wars,
coming to terms with it proves to be no greater challenge for the heroes
than had the "little wars" of imperial policing.

5.   The Great War: A War Like All the Others?

The Great War was an industrialised war. It was dominated by
machinery and high explosives, and became characterized by mass armies,
trench warfare, battles of attrition and high casualties.[51] It was perceived
by many as a new kind of war, much more terrible than all previous
conflicts.[52] The Western Front was and is still seen as the ultimate symbol
of this war, which was quite unlike the heroic and picturesque little wars
that had been depicted in pre-wartime writings - especially in children's
novels. To quote Samuel Hynes: "It was technology that had made the war
different... The guns of the artillery were bigger...; gas changed the odds
on infantry attacks; machine guns made massed assaults impossibly costly
in lives."[53]

It has frequently been claimed that especially in the years after
1916 (the year the battles of Verdun and the Somme took place) that a
process of disenchantment set in.[54] Instead of being perceived as an
exciting adventure, war was now seen as terrible and futile. This
disenchantment was largely due to the nature of the war on the Western
Front, which was static trench warfare with little movement or change and
little room for individual heroism. "War has never been looked at in the
same way since. It was no longer about personal fulfilment or honour or
even heroism. It was about survival."[55] The new attitude towards war
found its famous expression in the so-called "literature of
disenchantment," which in the war years included the war-poetry of
Siegfried Sassoon, Wilfred Owen and Robert Graves, and afterwards the
novels of Erich Maria Remarque and Richard Aldington.[56]

In recent years, however, this view of a general and far-reaching
disillusionment, and the accompanying shift from a romantic war
literature, which celebrated war, to a realistic one, which showed its "true"
and horrible nature, has been increasingly questioned. Rosa Maria Bracco,
for example, has examined works by so called "middlebrow writers" who,
rather than stressing the futility of the war, continued to emphasize the

justice and righteousness of the British cause.[57] Children's literature likewise displays no signs of disillusionment with war, either. While this is perhaps less surprising with children's novels (as the disenchanted war literature in the vein of Sassoon might have been considered rather unsuitable for children), the fact that the Great War is depicted as only marginally different from the colonial wars of the Victorian era is nonetheless telling. In fact, children's writers seem to take great pains to play down what was perceived as the special and different nature of this industrialized and mechanized "modern" war. Instead, they continued to paint war in a romantic and heroic light.

The most obvious result of this attempt is that few of the children's novels are set on the Western Front. Despite the fact that this was the most important theatre of war, which consumed more soldiers and produced more casualties than any other front, many of the novels do not describe it. Instead, they are devoted to the war of the so-called "side-shows": Africa, Mesopotamia, the Dardanelles and Serbia.[58] This is clearly shown by titles such as *How We Baffled the Germans: The Exciting Adventures of Two Boys In South-West Africa*; *Frank Forester: A Story of the Dardanelles*; and *Across the Cameroons: A Story of War and Adventure*.[59] War in these places seems traditional compared to the Western Front. It is a war of movement and strategy which leaves room for exciting cavalry charges and decisive action. One of the officers in Brereton's *On the Road to Bagdad* (1916) exclaims after an attack: "If that's war - the sort of war we're likely to have in Mesopotamia - then the more of it we have the merrier."[60] In these theatres, British soldiers faced well-known enemies, such as the Arabs, the Turks and the Boers. These are often shown to fight with rather traditional weapons; for instance, in Westerman's *Under the White Ensign* (1918) the attacking Arabs carry swords and spears. Clearly in such cases the war described is not at all like the anonymous mechanized nightmare of the Western Front.[61] This difference is explicitly stressed by the novelists, as can be seen in the following quotation from Brereton's *On the Road to Bagdad*:

> This desert warfare was so different from that which had now fallen upon the armies battling in Flanders against the Germans... Troops, both British and Turkish, were not sufficiently numerous to man a line running right across the country, and thus there was an opportunity to manoeuvre, the chance of out-flanking an enemy, and every now and again an opening for a charge, often brilliantly executed, by the British.[62]

Why writers should have preferred to describe this kind of warfare is obvious: it was much easier to keep the support and even enthusiasm for such a war than it would have been for the kind of warfare described by the "literature of disillusionment," which consisted of inaction and boredom, making soldiers victims and heroism impossible.[63] The war in Africa and the Middle East - at least as it was described by the novels - still provided the opportunities for individual heroism and initiative. In Herbert Strang's *Tom Willoughby's Scouts: A Story of the War in German East Africa* (1918) Tom wages a kind of guerrilla war against the Germans.[64] Though he is only supported by a few African men - none of them soldiers - he succeeds in harassing the German army to such an extent that it is no longer possible for them to attack the neighbouring British colony. While it seems difficult to imagine this kind of individual heroism in the trenches of the Western Front, the novels describing the action of the "side-shows" abound with it.

The war at sea also provides the setting for a number of novels.[65] Here the heroes experience exciting adventures in the Battles of the Dogger Bank and Heligoland, as well as patrolling the North Sea and catching German U-boats in order to stop that "inhuman crime, the submarine blockade of Britain."[66] It is perhaps in describing the war at sea that the novelists' attempt to downplay what was perceived as the specific nature of the Great War - industrialized, modern war dominated by technology - becomes most obvious. Rather, novelists continued to emphasize the importance of human qualities and the British naval tradition. In some instances, the authors seem to complain that these are no longer important:

> Such readers who look for single-handed gallantry on
> the part of our hero are apt to be slightly disappointed as
> [sic] this over-true story progresses, for modern naval
> warfare does not lend itself to the spectacular one-man
> actions of the olden days, when a Britisher was always
> certain of winning in a fight because of sheer pluck and
> determination. This is a scientific age, and those with
> the keenest brains generally prove themselves
> victorious; the old hammer-and-tongs pluck counts for
> very little.[67]

Yet the plot of Shaw's *With Jellicoe in the North Sea* (1916), from which the above quotation was taken, belies this complaint. Hal, the hero of the novel, is in fact able to achieve great things. In the Battle of the Dogger Bank, for instance, he single-handedly sinks a German destroyer. Later he and his crew mount a landing party in order to destroy a munitions factory in German occupied territory. Because of these deeds, he is frequently compared to the great Nelson himself. "Um - er - it's a bit young. And yet Nelson was a post-captain at twenty-one. I suppose there'll be lots of jealousy, but he's earned it."[68] The same comparison is made frequently in the case of Buckle in Walker's *Buckle of Submarine V2* (1916). In Nelson-like fashion, Buckle sinks parts of the German fleet in Kiel Harbour. Like the seamen of old he boards a German ship, upon which the narrator remarks: "It was as though history had rolled back a century and a quarter to the glorious days of the 'Wooden Walls,' when young Horatio Nelson himself had led his boarders against the *San Josef* and the great *San Nicolas*."[69] The aim of this comparison and the frequent references to the heroic past is to assure the reader that in spite of the changes in technology, the British will win again. Human qualities - as is even more obvious in the next quotation - will prove more important and thus British children have nothing to fear:

> Yet, could the great admiral come down from his statue
> in Trafalgar Square, and he tread the quarter-deck once
> more, he would assuredly say:
>   "My ships have changed, but my men are the
> same!"[70]

Claims such as this were another means of keeping up the spirit of the readership, which was not to be frightened by the pictures of helpless humans pitted against invincible technology that are normally associated with the First World War.

Ultimately, even those novels set on the Western Front attempt to give the impression that things have not really changed and that war is still that of the idealized past. The Western Front is presented very differently from what we see in the disenchanted perspective of the most famous war poets. The heroes of novels such as Lynn's *Oliver Hastings, V.C.* or Westerman's *The Dispatch-Riders* (1915) usually spend very little time in the trenches, and therefore do not experience the boredom and apparent pointlessness of trench warfare commonly associated with this theatre. They often serve in the cavalry or as dispatch-riders, and hence their war is mobile and active.

The small amount of time that the heroes actually spend in the trenches usually coincides with the great battles, for instance Loos, the Somme and Neuve Chapelle.[71] Whilst the Battle of the Somme, with its enormous casualties on the British side, is usually seen as the epitome of the horrors of trench warfare, such negativity is not evident in these novels. *Under Haig in Flanders*, in particular, shows the Somme as a great success, which almost achieved the decisive break-through:

> Let us go further...by stating that the conflict raged throughout the months of July, August, and September, and, checked by bad weather in October, proceeded once more in November... Even to outline the advances made, to describe in the broadest detail the difficulties met with, the triumphs achieved, and the gallantry displayed by our men, would be to pack our pages...
>
> Preparations for a renewed offensive in 1917 caused the enemy to consider the position seriously. For the Allies had been within an ace of wrecking their line and breaking right through in those operations of 1916.[72]

This claim is supported by the experience of the novel's hero, Roger, who is able to achieve great things in the Battle of the Somme. On the first day of the battle, Roger and his comrades capture a German trench, an achievement that is due to Roger's dash and bravery. Because Roger is slightly wounded, he is sent back to England to recover. Some weeks later, he returns to France as the commander of a tank. Little is said of what had happened at the Somme while Roger was away, but because he was such a success, it is implied that the attack as a whole must have gone well, too. With the aid of the tank, Roger and his crew drive deep into enemy territory and thus help the British forces to capture parts of the German lines.[73] Again, his experience serves to illustrate the success of the battle as a whole (and since he is caught by the Germans and spends the next weeks escaping from captivity, little else is said about the battle). According to the novels, even the war on the Western Front is therefore an unmitigated success and the British are sure to win there and in the other theatres of war.

Even the end of the Great War did not change the positive attitude of British children's war novels towards the conflict, and there was still no sign of the disillusionment which could be perceived in some of the most famous war literature for adults. As they had done in the war years, children's novelists in the post-war period continued to emphasize the necessity of fighting the Great War and to blame the Germans for its outbreak. They still celebrated British heroism and fighting spirit and attempted to console their readers for the many sacrifices and victims that the war had caused.

Notes

[1] Tom Bevan, *Doing His "Bit": A Story of the Great War* (London: Thomas Nelson, 1917), 52.

[2] For a description and definition of the genre see Michael Paris, *Warrior Nation: Images of War in British Popular Culture, 1850-2000* (London: Reaktion Books, 2000), 51-54; and Emer O'Sullivan, *Friend and Foe: The Image of Germany and the Germans in British Children's Fiction from 1870 to the Present* (Tübingen: Narr, 1990), 11.

[3] Dorothea Flothow, "'Told in Gallant Storie': Bilder des Krieges in britischen Kinder- und Jugendromanen für Jungen und Mädchen, 1870-1939" (PhD diss., University of Tübingen, 2005).

[4] Escott Lynn, *In Khaki for the King: A Tale of the Great War* (London: E.P. Dutton, 1915), v.

[5] See Jeffrey Richards, "Popular Imperialism and the Image of the Army in Juvenile Literature," in *Popular Imperialism and the Military 1850-1950*, ed. John M. MacKenzie (Manchester: Manchester University Press, 1992), 81-108; William J. Reader, *At Duty's Call: A Study in Obsolete Patriotism* (Manchester: Manchester University Press, 1988).

[6] Cf. Colin C. Eldridge, *The Imperial Experience: From Carlyle to Foster* (Basingstoke: Macmillan, 1996), passim.

[7] Hew Strachan, "Introduction," in *The Oxford Illustrated History of the First World War*, ed. Hew Strachan (Oxford: Oxford University Press, 1998), 1-8.

[8] Arthur Marwick, *The Deluge: British Society and the First World War* (London: Macmillan, 1965), passim.

[9] The propaganda effort is described in Cate Haste, *Keep the Home Fires Burning: Propaganda in the First World War* (London: Allan Lane, 1977).

[10] Ian Hay, *The First Hundred Thousand: Being the Unofficial Chronicle of a Unit of "K (1)"* (Edinburgh: William Blakewood, 1918 [1915]); Rudyard Kipling, "Mary Postgate," in Rudyard Kipling, *War Stories and Poems*, ed. Andrew Rutherford (Oxford: Oxford University Press, 1999), 235-249.

[11] Peter Buitenhuis, *The Great War of Words: Literature as Propaganda 1914-18 and After* (Vancouver: University of British Columbia Press, 1987), 14.

[12] Stéphane Audoin-Rouzeau, "Kinder und Jugendliche," in *Enzyklopädie Erster Weltkrieg*, ed. Gerhard Hirschfeld/Gerd Krumeich/Irina Renz (Paderborn: Ferdinand Schöningh, 2003), 135-141.

[13] Audoin-Rouzeau, p. 139. These articles can be found in the various

children's magazines published in the war years, e.g. *Chums*, *The Captain* and *The Girl's Own Paper*.

[14] See Escott Lynn, *Oliver Hastings, V.C.: A Realistic Story of the Great War* (London: W. and R. Chambers, 1916); Herbert Strang, *Frank Forester: A Story of the Dardanelles* (London: Henry Frowde, 1915).

[15] Ethel Kenyon, *Pickles; A Red Cross Heroine* (London/Glasgow: Colliers' Clear-Type Press, 1916), chapters III-IV.

[16] Brenda Girvin, *Munition Mary* (London: Humphrey Milford, 1918), 43.

[17] Lynn, *Oliver Hastings, V.C.*

[18] Frederick S. Brereton, *Under Haig in Flanders: A Story of Vimy, Messines and Ypres* (London: Blackie, 1917), 280.

[19] Bessie Marchant, *A Girl Munition Worker: The Story of a Girl's Work During the Great War* (London: Blackie, 1916).

[20] Enemies of other nationalities, such as the Turks or the Austrians, are much less important in the novels; in fact, novelists frequently claim that they have also been forced into the war by the Germans.

[21] See for example Percy F.C. Westerman, *The Dispatch-Riders: The Adventures of Two British Motor-Cyclists in the Great War* (London: Blackie, 1915).

[22] John Keegan, *The First World War* (New York: Knopf, 1999), 39. Cf. John Horne and Alan Kramer, *German Atrocities, 1914: A History of Denial* (New Haven/London: Yale University Press, 2001).

[23] This claim is very common in the novels; see also Charles Gilson, *Submarine U93: A Tale of the Great War, of German Spies and Submarines, of Naval Warfare and all Manner of Adventure* (London: "The Boy's Own Paper" Office, 1916), 15f.

[24] Lynn, *In Khaki for the King*, p. 124.

[25] Dorothea Moore, *Wanted, an English Girl: The Adventures of an English Schoolgirl in Germany* (London: S.W. Partridge and Co., 1916), 382. For further examples see Charles Gilson, "Held by the Enemy: The Thrilling Adventures of a Motor Scout," in *The Captain*, 33 (April-Sept 1915): 239.

[26] Herbert Strang, *A Hero of Liège: A Story of the Great War* (London: Henry Frowde/Hodder and Stoughton, 1914), 111; Rowland Walker, *Oscar Darnby, V.C.: A Tale of the Great European War* (London: S.W. Partridge, 1916), 99, 109; Robert Leighton, *Dreadnoughts of the Dogger: A Story of War on the North Sea* (London/Melbourne: Ward, Lock and Co., 1916), 261.

[27] Walker, *Oscar Darnby, V.C.*, p. 27; Westerman, *The Dispatch-Riders*, p. 31; Brereton, *Under Haig in Flanders*, p. 195; cf. also Frederick S. Brereton, *On the Road to Bagdad: A Story of Townshend's Gallant*

Advance on the Tigris (London: Blackie, 1916), 38.

[28] Lynn, *In Khaki for the King*, p. 2; Walker, *Oscar Darnby, V.C.*, p. 20.

[29] Klaus Vondung, *Die Apokalypse in Deutschland* (München: DTV, 1988).

[30] Dieter Langewiesche, "Zum Wandel von Krieg und Kriegslegitimation in der Neuzeit," in *Journal of Modern European History*, 2 (2004): 5-26, 11.

[31] John Buchan, *Greenmantle* (London: Thomas Nelson, 1942 [1916]), 93; Brereton, *Under Haig in Flanders*, p. 127.

[32] Tom Bevan, *With Haig at the Front: A Story of the Great Fight* (London: Collins' Clear-Type Press, 1916), 24; Bevan, *Doing His "Bit"*, p. 66; Herbert Hayens, *Under Haig and Foch* (London/Glasgow: Collins' Clear-Type Press, 1919), chapter X, n.p.

[33] See for example Moore, *Wanted, an English Girl*, chapter XVII; Nellie Pollock, *More Belgian Playmates: Heroes Small - Heroes Tall: A Second Story of the Great European War* (London: Gay and Hancock, 1915), passim.

[34] Bevan, *With Haig at the Front*, p. 150.

[35] Pollock, p. 150.

[36] Bevan, *Doing His "Bit"*, p. 175.

[37] See Robert MacDonald, "A Poetics of War: Militarist Discourse in the British Empire, 1880-1918," *Mosaic*, 23 (1990): 17-35.

[38] James Anthony Mangan, *Athleticism in the Victorian and Edwardian Public School: The Emergence and Consolidation of an Educational Ideology* (Cambridge: Cambridge University Press, 1981); Henry Newbolt, "Vitai Lampada," in *The Island Race* (Oxford: Woodstock Books, 1995 [1898]), 81f.

[39] Cecil D. Eby, *The Road to Armageddon: The Martial Spirit in English Popular Literature 1870-1914* (Durham: Duke University Press, 1987), 87.

[40] MacDonald, "A Poetics of War," passim.

[41] May Wynne, *An English Girl in Serbia: The Story of a Great Adventure* (London: Collins' Clear-Type Press, 1916), 67.

[42] Robert Louis Stevenson, *Treasure Island* (Manchester: World International Publishing, 1989 [1883]), 101.

[43] George Alfred Henty, *The Dash for Khartoum: A Tale of the Nile Expedition* (Mill Hall: Preston/Speed Publications, 2000 [1891]).

[44] MacDonald, "A Poetics of War," passim.

[45] Charles Gilson, *The Lost Column: A Story of the Boxer Rebellion in China* (London: Henry Frowde/Hodder and Stoughton, 1909).

[46] Buchan, *Greenmantle*, chapters VII-IX.

[47] Henty, *The Dash for Khartoum*, p. 355; H.A. Vachell, *The Hill: A Romance of Friendship* (London: Albatross, 1948 [1905]), 225.

[48] For details see Paul Crook, *Darwinism, War and History: The Debate over the Biology of War from* The Origin of Species *to the First World War* (Cambridge: Cambridge University Press, 1994).

[49] Eby, *The Road to Armageddon*, p. 2.

[50] Brereton, *Under Haig in Flanders*, p. 50.

[51] Strachan, "Introduction," passim.

[52] It has often been pointed out that many of the characteristics which were perceived as new and different about the Great War were in fact developments already present in earlier wars; see Brian Bond, *War and Society in Europe 1870-1970* (Leicester: Leicester University Press, 1983), 24.

[53] Samuel Hynes, *The Soldier's Tale: Bearing Witness to Modern War* (Harmondsworth: Penguin, 1997), 56.

[54] Paul Fussell, *The Great War and Modern Memory* (New York: Oxford University Press, 1975).

[55] John M. Bourne, *Britain and the Great War 1914-1918* (London: Edward Arnold, 1989), 235.

[56] Samuel Hynes, *A War Imagined: The First World War and English Culture* (London: The Bodley Head, 1990), passim.

[57] Rosa Maria Bracco, *Merchants of Hope: British Middlebrow Writers and the First World War, 1919-1939* (Oxford: Berg, 1993).

[58] See Keith Simpson, "The British Soldier on the Western Front," in *Home Fires and Foreign Fields: British Social and Military Experience in the First World War*, ed. Peter H. Liddle (London: Brassey's Defence Publishers, 1985), 135-158.

[59] Eric Wood, *How We Baffled the Germans: The Exciting Adventures of Two Boys in South-West Africa* (London: Thomas Nelson, 1917); Strang, *Frank Forester*; Charles Gilson, *Across the Cameroons: A Story of War and Adventure* (London: Blackie, 1916).

[60] Brereton, *On the Road to Bagdad*, p. 75.

[61] Percy F.C. Westerman, *Under the White Ensign: A Naval Story of the Great War* (London/Glasgow: Blackie, 1918), 147.

[62] Brereton, *On the Road to Bagdad*, p. 354f.

[63] John Onions, *English Fiction and Drama of the Great War* (Basingstoke: Macmillan, 1990), 50-56.

[64] Herbert Strang, *Tom Willoughby's Scouts: A Story of the War in German East Africa* (London: Oxford University Press, 1918).

[65] Cf. Gilson, *Submarine U93*; and Leighton, *Dreadnoughts of the Dogger*.

[66] Rowland Walker, *Buckle of Submarine V2* (London: S.W. Partridge,

1916), 52.

[67] Frank H. Shaw, *With Jellicoe in the North Sea* (London: Cassell and Co, 1919 [1916]), 94.

[68] Ibid., p. 309.

[69] Walker, *Buckle of Submarine V2*, p. 137.

[70] Ibid., p. 36.

[71] See Bevan, *With Haig at the Front*, chapters XVI-XVII; and Bevan, *Doing His "Bit"*, chapter IX.

[72] Brereton, *Under Haig in Flanders*, p. 183f.

[73] Ibid., chapters VII-IX.

Through Comic Eyes: *Punch*, the British Army, and Pictorial Humour on the Western Front, 1914-1918

John C. Horn

Abstract: Is tragic language the only suitable linguistic form through which to communicate the events of the First World War? This essay uses cartoons from *Punch, or the London Charivari* to examine the representation of humour in the British army during the Great War. It studies three cartoons from *Punch* to explore the myriad roles that laughter served for the British Expeditionary Force (BEF) on the Western Front from 1914 to 1918. Cartoons yielded (and still yield) many levels of historical meaning, and by analysing *Punch*, a source not often addressed by historians, this essay attempts to remove laughter from the margins of First World War scholarship and emphasize that humour was integral to the war experience. By no means is this proposing a replacement for what has come to be a *deficient* scholarly analysis; rather, the following pages should be considered an attempt to humorously complicate the memory of the Great War. With such an approach, what is most important for the author and the reader of this project is whether or not we understand what it means to "get" a joke, especially as we endeavour to "get" the past.

Key Words: Humour and the First World War, British Army, Cartoons, Pictorial Humour, *Punch, or the London Charivari*, Narrative and Textual Theory

1.   Introducing Humour and War

The British army's attritional plight on the Western Front during the First World War is seldom remembered as inspiring laughter. More likely, our popular memory dwells on the tragic imagery of the war, such as the "sudden and horrific" loss of 60,000 British soldiers (including 20,000 killed) during the first day on the Somme.[1] The British public was not prepared for the butcher's bill of over 670,000 dead BEF soldiers and, consequently, some historians have even gone so far as to say that tragic language is the only suitable linguistic form with which to communicate the events of 1914-1918.[2] Geoff Dyer's *The Missing of the Somme* (2001) uses the following fictionally constructed quotation as a means of emphasizing that, in many ways, the literary representation of the First World War has been compressed into a single cliché: "terrified, I clawed the stinking mud as the bullets whistled round my head and shoulders and as I waited for death."[3] While tragic language does typify the genre of Great War cultural studies, humour has been discussed, or dismissed, in paragraphs and chapters from Paul Fussell to Niall Ferguson; laughter *always* finds itself on the margins of histories that deal with the Western Front.

This essay uses the graphic and literary humour of *Punch, or the London Charivari* to examine the representation of humour in the British army on the Western Front during the Great War. In doing so, it searches for what has thus far been marginalized by scholarship on the conflict. Due in large part to this scholarly neglect, the *Journal of European Studies* (2001) devoted an entire issue to the broad concepts surrounding humour and modern warfare. Editors Valerie Holman and Debra Kelly, in their introductory essay, state that "humour is both cohesive and divisive; it *occupies all points of the sliding scale* between affection and cruelty, wit and buffoonery, expression of the status quo and subversion" (my emphasis).[4] As the myriad storylines engendered by humour (as well as the past itself) are of particular interest, this essay scrutinizes the notion of humour's many places on the "sliding scale" as a means of articulating both the enigmatic nature of humour itself and, perhaps more importantly, the ambivalent purposes that laughter served for British soldiers during the war. In the broader context of pictorial humour and the First World War, cartoonists commented on everything from the military hierarchy of the BEF to women in war to the "total war" political climate of Great Britain from 1914 to 1918; however, given the brevity of this modest project the scope and content of laughter surrounding *Punch*, the British army, and the Western Front will necessarily be limited.

Consequently, the content - or *targets* - in this essay addresses the representation of the German enemy, the war itself, leave, the home front, and life in the trenches, while the form - or *themes* - of humour includes punning, incongruity, farce and black humour. Cartoons yielded (and still yield) many levels of historical meaning, and by analysing *Punch* - a source not often examined by historians - this essay attempts to remove laughter from the margins of First World War scholarship and complicate the conflict's memory by emphasizing that humour, in all its forms, was integral to the war experience. With such an approach, perhaps the most important issue for the author and the reader of this project is whether or not we understand what it means to "get it." What kind of laughter passed through the lips of the soldiers and civilians who revelled at *Punch*'s cartoons? Was it supportive of the war? Critical? Or can it even be qualitatively analysed?

Dyer argues that the linguistic and thematic conventions of First World War cultural histories have become more powerful than the original experience itself.[5] He contends that it is impossible to write about the Great War without using language mirroring the literary and cultural responses of Wilfred Owen and Siegfried Sassoon. This is, as it happens, exacerbated by the fact that - whether one reads with a casual or scholarly interest - it has become extremely difficult to learn about the Great War except through the filter of Paul Fussell's *The Great War and Modern Memory* (1975), a groundbreaking examination and collation of the conflict's dominant themes.[6] Fussell begins his work by acknowledging that, correctly or not, it is with the words of David Jones, Robert Graves, Siegfried Sassoon and Wilfred Owen that we have come to form the current idea of the "Great War."[7] As a result, Fussell, as well as the literature which has followed his lead, created a standard of scholarship whereby the memory of the war becomes represented in the present, rather than in the past, where it actually happened.

In times of war, humour, for some, is no laughing matter. Perhaps this is why, in times of peace, it may seem that the only morally appropriate way of writing about war is to show that its seriousness should never be forgotten. After all, the best known cultural representation of the First World War experience for the British, as well as Canadian, public is the legacy of the war poets, particularly Owen and Sassoon. Their work has done much to fashion both the suitable ethical and artistic response to the human tragedy of the Great War. Such poetry of disillusion, its images rooted in the popular imagination that is very much a part of our cultural heritage, is even symbolized each year in Britain and Canada by the selling of poppies and by commemoration services all across the country.[8] Keep in mind Dyer's purposeful cliché as this essay analyses the Western Front experience, and, perhaps more importantly, how recalling and interpreting such events tends to be shaped by our modern memory of 1914-1918. A soldier's recollection of Passchendaele, or pictures of 1 July 1916 on the Somme, or the graveyards near Verdun, or the memorial at Vimy Ridge all account for the truthful poignancy of clichéd, tragic responses to the war. It is dangerous, however, to rely solely on such reactions, as they supplant many other facets of the war, such as humour, and can simplify the complex events of the past.

In a similar vein to Dyer, this essay searches for what is not there, what is missing, and what remains to be said concerning the way the Great War is remembered. Laughter has been suppressed and humour has been marginalized. Undoubtedly, British soldiers on the Western Front were more familiar with the pithy pictorial humour of *Punch* than they were with Wilfred Owen. Owen, who died only days before the Armistice, was not published until 1921, and his work was not popularised until the 1930s. In contrast, *Punch*, even with a paper shortage in Britain, increased readership and circulation during the war.[9] By the middle of the war, letters from the front even pleaded to those on the home front to send publications other than *Punch* to the trenches, as the dugouts along the

Western Front were already festooned with what would come to be known as "the last word in English humour."[10] Laughter found itself amongst the mud and blood of the First World War, but the names of cartoonists like Bert Thomas, F.H. Townsend, and Dick German, whose works are the empirical focus of this essay, tend to be left out of the literature that deals with the conflict.

In many ways the literary conventions, typically shrouded in Fussellian discourse, of the scholarship surrounding the Great War account for the marginalisation of studies on humour; and no matter what stance historians take on the literary nature of the war, it seems that time and again arguments are immersed in tragic language. Nevertheless, this essay does not pretend to be a replacement for what is a *deficient* analysis of the Great War and its horror and tragedy. Rather, it simply suggests that along with the remembrance of the war's tragedy also goes the recollection and study of its humour. Standard historical approaches to the Great War yield much truth, but they also oversimplify an extremely complicated event in history by reducing it to only tragic elements. Studying the humour of 1914-1918 begins a process that clarifies both the ethical reaction to the war *at the time* and the complexity of the historical and cultural response that followed.

2.　　　　Complexities of Humour: Storylines of the Past

Two schools of thought seem to have emerged from the historiography of humour and the Great War. Laughter has been interpreted as both a critical and supportive voice towards the war, with most of the above works favouring the latter approach, as is evident in Denis Winter's comment that "the positive, supportive side of a static society, whose roles were simple, was a particular type of humour."[11] My analysis of *Punch*'s pictorial humour differs from previous scholarship in that it approaches laughter and war without trying to channel the complex concept into a definitive conclusion. For as we move to examine the abstract notion of humour, it will become perfectly muddled and muddily clear that the roles laughter occupies on the "sliding scale" do not fit so well into one-dimensional spaces.

Humour employs many vehicles, or themes, and everything in existence can be made the butt of a joke; however, the concept is not always similarly presented and such themes are seldom equally evident from one theory to the next.[12] Any theory of the comic must be judged by its consistency to itself and its faithfulness to the data it explains. It must untangle old problems and lead to new insight.[13] This essay employs four themes of humour to analyse funny material from the First World War: *puns*, from their lowest to most sophisticated forms, are vehicles of humour that facilitate examination of the wordplay between, among other things, cartoons and their captions;[14] *incongruity* deals with exaggeration, understatement, juxtaposition, mimicry, and coincidence/presupposition as a means of demonstrating the "lack of fit" between, for example, a joke's punch line and the reality of the First World War;[15] *farce*, including satire, sarcasm, bathos, and caricature, serves to ridicule and prove absurd the subject(s) of a joke; finally, *black humour* proves that "nothing is so sacred, so taboo, or so disgusting that it cannot be the subject of humour."[16]

It is necessary to dwell briefly on the technical aspects of cartoons, as it is through the humour of these cultural objects, these simplistically complex scribblings, that scholars can find their window into the comic world of the Western Front. L.A. Doust sees cartoons as being similar to puns, in that their many levels of meaning range from simple sketches to complex works of art involving several characters; furthermore, while important, the art, or form, of a cartoon is not as vital to its humorous success as is the joke, or the drawing's content.[17] *Punch* cartoonist Cyril Kenneth Bird, known by the pseudonym "Fougasse," even referred to himself as a "pictorial humourist," if only "to stress the fact that the *humour* is more important than the *art*" (my emphasis) and that cartoons should be judged by their humour rather than by their art.[18] Critics and scholars seem to have reached a consensus regarding the belief that styles of drawing should be adapted to types of humour (such as pun, incongruity, farce, and black humour), and vice-versa. These concepts will become clearer as the cartoons are analysed. Much like humour, this essay endeavours to achieve new meaning with its conceptual approach to the past and, specifically, to the First World War.

Employing multiple viewpoints (both supportive and critical) towards *Punch*'s pictorial humour of the First World War emphasizes the fact that cartoons often had many levels of meaning. Be wary! Clearly historical pitfalls (or "shell-holes") are everywhere in this discussion of the Great War. Writers of the conflict (perhaps pictorial humourists as well) often sought to let the "facts" speak for themselves, descriptively expanding their historical narration with the intention of creating a strong illusion of reality. However, by grounding a work of literature, whether it is fictional or not, in the historically "real" does not necessarily make it real. As this is an essay about laughter, perhaps a joke can help explain the problems of communicating with the past, especially when examining the cultural representation of graphic humour:

> Two Englishmen are riding on a noisy train. The first
> says, "I say, is this Wembley?" The other responds, "No,
> Thursday." To which the first says, "I am too."[19]

The two men have completely misunderstood each other, but no matter -
they both *thought* they were communicating. Such is the job of the
historian. We endeavour to bridge this gap in comprehension. We find
what has been marginalized, what has long been suppressed, perhaps even
what was thought to be lost. Laughter presents another facet of the war
experience, and through humour scholars can begin to explore yet another
suitable storyline with which to remember the Western Front of 1914-
1918.

    Moreover, as "narrative structure is not arbitrarily imposed on the
events of the past, but inherent to them," it is important to recognize the
many possible facets of *any* historical text, such as a cartoon.[20] This
essay's approach to humour and the Great War is paradoxical. It both
challenges and reinforces the themes of the conflict by acknowledging the
simultaneously supportive and critical nature of *Punch*'s pictorial
humour.[21] By recognizing pictorial humour's, as well as the past's,
contradictions and their relationship to the "sliding scale" this essay
endeavours to achieve a richer, multi-dimensional understanding of
history.

3.   *Punch*lines

    From what we know, *Punch, or the London Charivari* was
founded in either 1840 or 1841 by Joseph Last, the printer, Ebenezer
Landell, the engraver, and Henry Mayhew, the playwright and journalist;
but even today the origins of the magazine are still debated in the United
Kingdom. During his first years, Mr. Punch expressed his "reputation for
being a defender of the oppressed and a radical scourge of all authority"
by speaking out against the imperial adventures of the British Empire as
well as continental despotism, which subsequently saw the magazine
banned in France and Austria.[22] After 1850, however, Mr. Punch
increasingly reflected the more conservative views of Britain's growing
middle class, which was the publication's primary readership
demographic.[23] By 1914 over 100,000 issues of *Punch* were being
distributed each week in Britain, and with cartoonists like Thomas,
Townsend, German and Fougasse enduring the trenches for "the last word
in English humour" the nation's obsession with the magazine began to
grow. By the end of the war, and in spite of the paper shortage, *Punch*'s
circulation reached 200,000 issues per week. Undoubtedly, this came
about because of the intense patriotic fervour generated by the magazine.[24]

With technological innovations in transportation and cartooning (specifically the transition from the woodblock to photographic printing), information, and in this case pictorial humour, had never moved so quickly between home front and the war front. But did this result in Mr. Punch's influence reaching an extent never before thought possible?

Even though the shipping of cartoons between London and the Western Front was quite rapid, it is important to clarify what has been alluded to several times in this essay: cartoons represented the essence of a certain event or person from the First World War. Though specific occurrences and people were made the subject of jokes, pictorial humour from the Great War seems to have served the larger purpose of commenting on the greater, symbolic meaning of the conflict. Moreover, while humour helped to construct a national identity during the war, this patriotic construct was often fractured along class, gender and civilian-military lines. Political and social cartoons even engaged these divisions and, consequently, the joke would have to be presented such that all parties implicated in the laughing process would "get it."[25] Often relying on pre-war values and assumptions, "the mechanism of humour is certainly very complex, but in the case of the pictorial humour displayed in the popular press between 1914 and 1918, [Purseigle] would contend that the mediation of war experience it offered relied on the lowest cultural denominator."[26] Once again we see that humourists made sure that their work could, ideally, be understood by anyone who picked it up; however, even though catering to the lowest cultural denominator made sure that everyone "got it," such an approach did nothing to ensure that soldiers, civilians, politicians, and historians all "got it" in the same way.

4.   Introducing the Cartoons and Cartoonists

Before undertaking analyses of this essay's three selections from Mr. Punch's cartoons, some introductions must be made. Dick German, who ironically refused to change his name or adopt a *nom de plume* during the war, published just four cartoons for *Punch* - two in 1914 and two more in 1915. He also contributed cartoons to the South Wales Daily Post published in Swansea between 1914 and 1916, the year of his death on the Western Front. He created his 20 January 1915 cartoon (appendix, Figure 1) in an effort to depict the jovial nature of Mr. Tommy Atkins - even when the archetypal British soldier finds himself outnumbered three-to-one by German enemies. The farcical punch-line of German's cartoon involves an unarmed, cheerful Tommy confronting three "Jerries," stereotypically attired in *pickelhaube* helmets. The Germans appear utterly confounded by the polite effrontery of their foe and, though they are fully-armed, seem to be no match for the chirpy Tommy.[27]

F.H. Townsend, the "archetypal *Punch* cartoonist of the
Edwardian era," contributed to *Punch* from 1903 and became the
magazine's first art editor in 1905. He was born on 25 February 1868 in
London and studied at the Lambeth School of Art. During the war he
served in the Special Constabulary at home in England.[28] Townsend,
assisted by Bernard Partridge and Leonard Raven-Hill, oversaw the
production of cartoons until 1920, when he died while playing golf, a
rather fitting end for a true *Punch* man. In "The Irrepressibles" (appendix,
Figure 2), a cartoon from 20 February 1918, Townsend seems to be
arguing that sceptical laughter was a perfect means with which to endure
and criticize the circumstances on the Western Front. This sort of cynical
humour is typical in Townsend's contributions. In comparison with the
generally supportive 1915 cartoon by Dick German, "The Irrepressibles"
shows the changing tone of *Punch*'s humour regarding the plight of
Tommy. It also plays on the term "irrepressibles," an age-old
colloquialism referring to the unstoppable nature of British soldiers.

Born in Newport, Wales, in 1883, Bert Thomas brought his style
to *Punch* in 1905 and contributed to the magazine until 1935. During the
First World War, Thomas served as a member of the Artists' Rifles and
became nationally known for his cartoon "'Arf a mo', Kaiser" of 1914,
which he drew as part of a £250,000 promotion to supply troops on the
front lines with tobacco and cigarettes.[29] Thomas was also a member of
the London Sketch Club, founded in 1898 by a group of Fleet Street
draughtsmen. The organisation was made up of several young actors and
writers, but it was artists like Thomas who were "busily reinventing an
entirely new form of pictorial humour, as a relaxation from illustrating the
classics."[30] Thomas's cartoon from 2 January 1918 (appendix, Figure 3),
addresses the inability of the home front, especially from the collective
viewpoint of the soldiers in France, to grasp the reality of the Western
Front. Cartoonists such as Thomas represented a minority viewpoint of the
media's "reproduction" of the war experience, as he actually found
himself on the front lines of the Western Front. This could not be said of
many journalists, artists, or humourists, such as Townsend and German;
nevertheless, one must credit the staff at *Punch* for approaching the Great
War with just the right amount of scepticism regarding the information
that was being received from France.

Finally, the playful figure in all three cartoons, Mr. Tommy
Atkins, must be placed in a context. In 1812 a War Office publication
showing the proper way to organize the *Soldier's Pocket Book* gave as its
example one Private Thomas Atkins, No. 6 Troop, 6th Dragoons. During
the First World War the nickname was widespread. "Tommyness" was
used to describe certain attitudes and behaviour, and talking "Tommy"

defined the congenial repartee between soldiers.[31] Indeed, Tommy *was* the British soldier, but one must keep in mind that he also wore *many* guises, depending on who was depicting him, his dialogue, and his actions. Through these three cartoons this essay investigates the myriad possibilities of being Tommy Atkins.

5.   The Cartoons: Simultaneously Supportive *and* Critical

As an audience, how do we "get" the three cartoons documented in Figures 1-3? Truth be told, the war did much to expose the enigmatic nature of *Punch*. Laughter occupies all points of the "sliding scale," and so too did Mr. Punch and his staff of pithy humourists. They spoke out against *both* subversion and the status quo. Consequently, it is necessary to establish some boundaries for this essay, to point out the critical and supportive roles of laughter during the First World War and explain how these particular cartoons relate to the "sliding scale."

Laughter, a hilarious critique of the First World War: It did not take long for the soldiers of the Great War to realize that twentieth-century weapons combined with a strategy of attrition spelled disaster. It was the humourists, with their jaded and irreverent perception of the Western Front, who provided a cynical solution to the patriotic rhetoric regarding the heroic nature of war. A 1916 *Punch* cartoon, which satirically depicts "the cheerful one" (a veteran of the Western Front) giving words of advice to a newcomer in the trenches, does much to trivialize the grim reality of the Western Front: "If yer stands up yer get sniped; if yer keeps down yer gets drowned; if yer moves about yer get shelled; and if yer stands still yer gets court-martialled for frost-bite."[32] The creative minds of *Punch* were very effective in satirizing the British army to the point of mocking the heroic representation of battle and made laughter an integral part of the war experience.[33] Contributors to these works saw their collective pen as being mightier than the sword, especially when those whom they were criticizing, such as Lord Kitchener or Sir Douglas Haig, could, realistically, only be attacked with pen-strokes. And attack they did, as cartoonists criticized everything from recruitment to officers to the conditions of the Western Front.

Laughter, a weapon of the British army: If you can't beat 'em, join 'em! Recognizing that laughter, even if it was at the expense of the war effort, was an integral and unchangeable part of the war experience, the British government mobilized humour as a weapon of war. Humour was transformed into an abstract value representative of domestic, common traits and made to serve the war effort and national sentiment. Therefore some of the comedic works discussed below should be interpreted "as a means of inspiration for loyal service and for the cheerful endurance of hardship...for all groups of the English race now [had] their boys and their hearts engaged in this great struggle."[34] Laughter of the British sort was something worth fighting for, not only behind the lines but also at the front, where military authority let caricaturing continue because it gave soldiers a chance to relax and distance themselves from the war's horrific reality.[35] In many different ways humour motivated the soldiers of the BEF to keep fighting. Laughter was simultaneously critical and supportive of the conflict, and, consequently, the following selections of pictorial humour must be analysed both on their own and as parts of the broader context of the Great War. By doing so, readers will begin to properly explore the interconnected discursive differences of *Punch*'s war cartoons.

Let us begin where so many British humourists began in the summer of 1914: with the enemy. The German (Boche, Fritz, Hun, Jerry, Kraut) was satirized in countless ways from 1914 to 1918 in an effort to motivate and inspire the British war effort, both at home and in the trenches. Exaggerating aspects of a person or object as a means of fleshing out certain features is one of the multiple functions of humour.[36] By dwelling on stereotypical German traits or characteristics, portraying the enemy as faceless, or alternatively by showing Germans to possess only evil or, perhaps, clumsy attributes, cartoons did much to motivate soldiers to engage their adversary. Whatever the context of Tommy encountering Fritz (or vice-versa) in a *Punch* cartoon, the British soldier *always* won the joke over the German. The enemy wore many humorous guises, all of which tended to be supportive of the British war effort, thus reflecting the positive meaning of this particular cartoon.

Pierre Purseigle argues that laughter allowed the soldiers of the First World War to maintain control over their experience, as pictorial humour drew a joyful veil over the grim reality of the Western Front.[37] Robert Graves commented on the comedic juxtaposition of the prevailing opinions amongst soldiers in the trenches: "we held two irreconcilable beliefs: that the war would never end and that we would win it."[38] It was this grim optimism that Tommy seems to be taking to the battlefield in Figure 1, a pre-Somme and Passchendaele caricature.[39] The wording of the caption is rather positive, given the fact that the Germans seem to have just expelled a British force from a trench line; moreover, the word "temporarily" suggests that the BEF's counterattack will land them back in the lost trench, perhaps for good this time. Presupposition makes the joke funny, as the audience does not expect Tommy to be so playfully polite to the enemy. The cartoon supports the concept that "there is no kinder creature than the average Tommy. He makes a friend of any stray animal...When he's gone over the top...for the express purpose of doing in the Hun he makes a comrade of the Fritzie he captures."[40] Whether kindness will be reciprocated by the Germans is a notion that is left up to the cartoon's audience.

Figure 1 also incongruously violates logic, and, by employing farce, can do nothing but distance soldiers from reality, as the audience is inspired to believe that the Germans' bullets will simply bounce off the chirpy Tommy.[41] The British soldier depicted in the scene is laughing in the face of his would-be executioners, thus emphasizing the cartoon's gallows humour. Given Charles R. Gruner's notion of "aggressive humour," in which a figure of a caricature is robbed of any superiority, the cartoon serves to inspire morale: Tommy wins the joke, as all he needs to combat his enemies is a smile, a glib remark, and, in this particular instance, his pipe.[42] This *Punch* cartoon ridicules the German enemy to the point of making Tommy the joke's superior subject. Alternatively, perhaps such a representation of a soldier's material desires humorously defuses what would, in reality, have been a fearful encounter with German soldiers. Is the punch-line of Figure 1 somewhat skewed? Corporal H. Diffey, 15th Battalion Royal Welch Fusiliers, commented on a different sort of laughter engendered by the supportive humour of Figure 1: "we could laugh out loud at these reports, plagued by lice and living amongst the debris of war and the legends that sustained our armchair patriots at home."[43] Once again the ambiguity of laughter on the Western Front is revealed.

To "get it": "Any one who talks of the glory of war should be invited to walk over a battlefield when the fighting has ceased. He will see those who have 'got it' from shells or bullets writhing in agony, he will hear many of them asking someone for the love of God to kill them and put them out of their misery."[44] Here was a different way to "get it" in the First World War. Regardless, the idea of "getting" a joke and "getting" shrapnel stemmed from the same concept, which emphasized the wholehearted un-glorified nature of modern warfare.[45] According to John Keegan, by the end of the war the odds of individual survival had passed from the *possibility* to the *probability* of death for the British soldier.[46] In Figure 2, Tommy Atkins does not appear to be as chirpy as he was in the pre-Somme, pre-Passchendaele Figure 1.[47] The term "irrepressible" is given new meaning here, as Tommy is, in fact, stopped and repressed by a strand of barbed wire. Moreover, Fred W. Leigh's 1915 song, "The Army of To-Day's Alright," is debased, both by Tommy's cynical statement as well as the scene depicted by F.H. Townsend, in which at least one of the advancing soldiers has been taken down. The ballad's patriotic call to arms ("Boys, take my lip and join the army right away") and its glorification of life in the BEF (the "Army's simply perfect") seems to

contradict the scene in Townsend's cartoon.[48] Townsend's army appears
to be bent, tangled, and nearly broken, rather than the "perfect" collection
of "big strong chaps" described by Leigh's lyrics. This being said, scholars
like Gary Sheffield contend that there is actually a great deal of truth to the
music-hall comedian's remark as well as the title of Figure 2. By 1918 the
British army "was at the peak of a learning curve."[49] Further, the
"irrepressible" attack during the last 100 days of 1918, according to
Captain D.V. Kelly, 6th Leicesters, 21st Division (also a veteran of the
Somme), "gave striking proof of the enormous advance made by the new
British Army in the technique of warfare, for it was a small masterpiece
achieved with one tenth of the casualties it would assuredly have cost us in
1916."[50]

    The BEF's final offensive made it impossible for the defenders to
respond, and, perhaps more importantly, it was carried out at the pace of
the poor bloodied infantry. Even at the end of the war, the BEF could only
move as fast as its slowest foot soldier, which makes the content of
Townsend's cartoon all the more relevant. Perhaps there is some truth to
Leigh's claim in the song, which is from the perspective of a soldier, that
"I joined the army yesterday, so the army of to-day's alright!"[51] It was,
after all, the "irrepressibles" who dictated the speed of the unstoppable
attack.[52]

    The presupposition of Figure 2 stems from the incongruity
between the cartoon and its caption, as the pictorial humour does not seem
to fit with the notion of Tommy being a force that is impossible to repress
or control. Townsend's work satirizes the home front for, once again,
being out of touch with the soldiers at the battlefront. Not only does the
home front not comprehend the plight of Tommy, but somewhere in
Britain there is a music-hall comedian making "three hundred quid a
week" by trivializing the horrors of attritional warfare on the Western
Front. This is suppressed by the portrayal of Tommy tangled in barbed-
wire and face down in mud. Clearly there is a lack of fit between this sort
of statement and the bleak reality of the Western Front, as Tommy is not
heroic and the trenches are not at all romanticized in this cartoon. The
black humour in Figure 2 proves that even the concept of a fallen soldier
was not so taboo that it could not be discussed in *Punch*.

    Evidently some humourists recognized that the British army was
experiencing trying circumstances as it struggled to destroy the German
force once and for all. With the advantage of hindsight on our side,
contemporary scholars know that things would only get worse on the
Western Front for the BEF in the months that followed the publication of
Figure 2, as the Germans would go on their last major offensive. Journalist
Horatio Bottomley (the self-proclaimed "soldier's friend') and his work

"Somewhere in Hell" were met with unbridled cynicism and, in some cases, sarcasm-ridden anger from the men on the front lines.[53] By this point in the conflict few soldiers had much patience for the romanticized insight of the home front. They *were* in hell, and no journalist, perhaps even no soldier, could communicate this phenomenon accurately.

Thomas's work, Figure 3, uses punning, specifically language humour, as the Scottish artist queries to Tommy with his thick accent regarding the realism of the scene to be painted: "Ah dinna ken what it is. It doesna seem realistic enough."[54] The dialectic exchange is made the subject of humour, and if the language used by the artist is not enough for archetypal categorisation, then Thomas's portrayal of the painter in a stereotypically Scottish hat does the job. The accessories a pictorial humourist provides his cartoon characters are also of vital importance, as the comic addition of riding pants, a rifle, or hat can, at times, tell more about the figure than the caricaturing of his or her face and features.[55] Perhaps Tommy being attired in full military garb, even while home on leave, suggests that he can never, and will never, escape the Great War. Even back in Britain he is being made to "go over the top." Moreover, what does it mean when Tommy claims that the only accessory missing from the scene is his "tot o' rum"? The dialogue in the cartoon drips of sarcasm, especially when one takes into account the artistry of the cartoon. The words "have we forgotten anything?" are humorously juxtaposed with the artistic content of the cartoon, which contains paintings and pictures, a flower vase, dressers, and, perhaps most importantly, a wooden *construction* of a trench. This serves to articulate the absurd notion of accurately reproducing events of the Western Front. The fallacious representation of the trench is steeped in just as much inaccuracy as the Scottish artist's work.

Several incongruous features riddle the cartoon. Tommy understates the absurdity of attempting to capture the essence of the Western Front in a studio somewhere in Britain. He suggests that the only lack of fit between reality and the scene to be painted revolves around his being without the "tot o' rum" that was occasionally given to troops before going over the top. The concept of the "tot o' rum" is itself steeped in humour. First, this shot of rum should not be thought of as the "domestic apology" served as rum today: "burning the throat, it half choked the soldier and made his eyes tear violently."[56] Rum was usually contained in a one gallon earthenware jar, which was supposed to be enough for 64 men. The letters "SRD" on the jar arguably stood for "Services Rum Diluted" or "Special Rum Distribution"; however, the soldiers invented new apronyms, such as "Soon Runs Dry" and "Seldom Reaches Destination."[57] Corporal J.G. Mortimer's 10th Battalion, The York and

Lancaster Regiment, commented that Tommies often went into battle "with the best of luck and a spoonful of rum."[58]

Phillip Gibbs described the trenches as being "alive with a multitude of swimming frogs"; the sides of the trenches were covered with slugs and horned beetles; and rats were among the worst horrors: they fed on corpses, grew to be as large as cats, and sometimes even attacked sleeping men.[59] This all seems to be understated in the civilian artist's take on the war, but is emphasized in other *Punch* cartoons, particularly Figure 2, which even shows a frog fleeing from a tumbling Tommy. Was this life on the Western Front? The dangers of the trenches and an outright critical account of the war (one that is not tongue-in-cheek, at least) seem to be ironically marginalized by Thomas's work. Perhaps this suggests the supportive nature of the cartoon.

As supportive as the pictorial humour of Figure 3 may be, one must consider the response of what a cynic, like Corporal H. Diffey, might call a "romanticized" or "sanitized" spin on the war experience. By taking a critical line, the audience of Figure 3 can rightfully discern that Tommies along the Western Front saw their grim reality made up of much more than a "tot o' rum." Given this, we can interpret the cartoon as Thomas's satirical representation of many aspects of the British media regarding their approach to the war from home. Due to the fact that he, unlike other figures in the British print-media, fought on the Western Front (as a member of the Artists' Rifles), Thomas probably saw himself as being well-suited to accurately produce material pertaining to the Great War.[60] Figure 3 offers a sceptical commentary on the reality of the war experience by comically juxtaposing a *real* Tommy with a reproduction that is an absurd farce. Perhaps this is Thomas's comment on the difficulty of communicating what is "historically 'real' but unnameable."[61] Though he is a survivor of the Western Front, Figure 3's Tommy has a troubling time recounting his war experience to the Scottish artist. The war experience has become unnameable. Tommy, hopelessly misplaced in a trench somewhere along the Western Front, cannot even calculate time. It seems that even the true soldier cannot always be truthful, even if he wants to be.   Unable to withdraw from the experience of the Western Front, Thomas's soldier finds himself accessorized with full military attire and in search of his "tot o' rum" even though he is hundreds of kilometres removed from the trenches. Perhaps this suggests that the "real" Tommy will never get away from the horrors of war, thus revealing the critical side of the cartoon, and the war experience in general. The complex, perhaps indecipherable, reality of the Western Front will forever weigh on the heart and mind of Mr. Tommy Atkins.

6.   The Un-Marginalisation of Humour and War

Cartoons and humour evidently had the power to persuade the population, something which the British Government recognized and, consequently, endeavoured to harness for its benefit.[62] Developments in culture and cultural diffusion saw humour extend and maintain clout the likes of which it never had before, a trend that continues today, and which still operates along a "sliding scale."[63] Laughter provided a coping mechanism for the bleakness of the First World War, and the idealized characters of Tommy and Mr. Punch helped to distance soldiers from reality. This conception of humour argues that laughing resulted in the release of nervous energy, thus allowing Tommy to be a better soldier.[64] In 1935 Stephen Leacock pointed out that "one can at any rate find a certain distinctive individuality that marks the humour of each country. No doubt as the nations unify and standardize out of existence the distinction will tend to fade."[65] During the First World War humour of the English sort had not yet begun to fade. As a result, whether cynical or supportive, it also represented a part of the English national character, something worth fighting to preserve. Humorous cartoons and literature held a mirror, albeit a hilariously distorted one, to the trench experience during the Great War; they therefore allow scholars to use laughter as an entry point into the enigmatic world of the Western Front of 1914-1918.

This essay has balanced on the fine line between removing humour from the margins of Great War cultural scholarship and the trivialisation of war itself, and has done so purposefully. Has this work made light of what has become a tragic genre of historical study? Perhaps, but after reading this essay at least *one* person will, hopefully, think differently of the Great War and its modern memory. "We have traversed the gamut of sensation," said Mr. Punch in 1917, "from the sublime and tragic to the ridiculous."[66] From the humorous horrors of the trenches to the absurdity of reproducing the Western Front in a British studio, this essay has aimed to take its reader on such a journey. By embracing humour's, as well as the past's, "sliding scale," multiple meanings can be reached, depending on how an author uses acts of narrative to analyse the material created by those who experienced the events of the past. For scholars, different stories can plot multiple courses through the same archives, libraries, and rare-book rooms of the historian's world. Does such a concept blur the notion of *reproducing* and *producing* the war experience?

It would seem that the first-hand knowledge of war writers gave them credibility regarding the reproduction of the events of 1914-1918 on the Western Front: "when experience is taken as the origin of knowledge, the vision of the individual subject (the person who had the experience or the historian who recounts it) becomes the bedrock of evidence on which explanation is built."[67] When such a statement is juxtaposed with the following comment by renowned war poet Robert Graves, the credibility of the war authors is somewhat brought into question:

> The memoirs of a man who went through some of the worst experiences of trench warfare are not truthful if they do not contain a high proportion of falsities. High-explosive barrages will make a temporary liar or visionary of anyone; the old trench-mind is at work in all over-estimation of casualties, "unnecessary" dwelling on horrors, mixing of dates and confusion between trench rumours and scenes actually witnessed.[68]

Can, as Wilfred Owen once said, true poets really be truthful? Can historians? As scholars of the past, all that we can do is offer the idea that no matter how documentary, factual or autobiographical an account of the Great War presents itself, it is just a war story among other war stories: "if reality remains inaccessible or unnameable," says Evelyn Cobley, "then all narrative renderings produce rather than reproduce the war experience."[69] Perhaps Thomas's work, which perfectly supplements Graves's analysis, foreshadowed such an idea, as it comments on the absurdity surrounding the "real" reproduction of the war experience.

    The works analysed in this essay are representations, as well as constructions, of cultural meaning during the First World War. The supportive and critical authorial voices of this essay have expanded such meaning beyond what was, probably, the intention of the humorous individuals who created something worth examining today. Such is the job of the historian. We endeavour to bridge this gap in comprehension. We find what has been marginalized, what has long been suppressed, perhaps even what was thought to be lost. Laughter, which has been shown as a vital part of the Great War, is not the first thing that we think of when we remember the conflict. It might not even register in our minds. As scholars of the past do we feel, rather than think, too much about the First World War? We should keep in mind that the elegies of Owen, Sassoon, Graves, Blunden and the other war poets were not even widely read until the 1930s, while, for whatever reason, humorous material was always very popular - perhaps none more so than *Punch*, "the last word in English humour." So, we find ourselves where we were at the beginning of this story, grappling with the enigma that was humour and the First World War. What kind(s) of laughter passed through Tommy's lips as he sat in a muddy trench, wrestling with a rat over the last tin of bully beef?

Undoubtedly, differing circumstances and personalities assured that the laughter echoing amongst the bursting shells or sniper fire occupied all points of the "sliding scale."

Perhaps the most telling piece of humorous material to survive the mud and blood of the First World War is a brief verse that encapsulates the fatalistic attitude of Tommy Atkins. Mr. Punch's Western Front correspondents noted the relevance of the verse in August 1915: "there is a deservedly popular military song which states, with perhaps unnecessary iteration, that the singers are there because they're there, because they're there. That is exactly how we find ourselves placed at the moment."[70] Tommy - the, at times, unwilling hero - even if he was fed up, would never give up. Even today, we can "get" this black humour, especially when considering the tragic elegies on which we have been brought up. War was not, and is not funny, but perhaps it is not always tragic either:

> We're here because we're here
> Because we're here
> Because we're here.

And humour was part of "being here," for the soldier as well as the historian.

Acknowledgements

The heart and soul of this essay are rooted, like any good piece of history, in fact. Luckily for me, the "facts" that I grubbed in the archives to find were hilarious and extremely fun to research. Such pleasure-filled giggles are rarely sparked by hours and hours of painstaking research, and for this I thank the humourists of *Punch, or the London Charivari*, who produced something well worth examining today. The cartoons in this essay have been reproduced with the permission of *Punch Ltd.* Many thanks to them for letting me use age-old British humour to put a modern spin on First World War cultural history.

Notes

[1] Gordon Corrigan, *Mud, Blood and Poppycock: Britain and the First World War* (London: Cassell, 2003), 272.
[2] Corrigan, 54.

[3] Geoff Dyer, *The Missing of the Somme* (London: Phoenix Press, 2001), 81.

[4] Valerie Holman and Debra Kelly, "War in the Twentieth Century: The Functioning of Humour in Cultural Representation," *Journal of European Studies* 31 (2001): 249.

[5] Dyer, 81.

[6] Ibid., 84.

[7] Paul Fussell, *The Great War and Modern Memory* (New York and London: Oxford University Press, 1975), 3.

[8] Holman and Kelly, 249-250.

[9] Jay Winter, *Sites of Memory, Sites of Mourning: The Great War in European Cultural History* (Cambridge: Cambridge University Press, 1995), 212-215.

[10] Stephen Leacock, *Humour: Its Theory and Technique with Examples and Samples: A Book of Discovery* (London: The Bodley Head, 1935), 217.

[11] Denis Winter, *Death's Men: Soldiers of the Great War* (London: Penguin Books, 1978), 233-234.

[12] James K. Feibleman, *In Praise of Comedy: A Study in Its Theory and Practice* (New York: Horizon Press, 1970), 203.

[13] Paul H. Grave, *Comedy in Space, Time, and the Imagination* (Chicago: Nelson-Hall, 1983), 32.

[14] Charles R. Gruner, *The Game of Humour: A Comprehensive Theory of Why We Laugh* (London: Transaction Publishers, 1997), 324.

[15] John Morreall, *Taking Laughter Seriously* (Albany: State University of New York Press, 1983), 77.

[16] Gruner, 44.

[17] Len A. Doust, *A Manual on Caricature and Cartoon Drawing* (London: F. Warne, 1945, 1932), ix; Fougasse, *The Good-Tempered Pencil* (Toronto: Max Reinhardt, 1956), 18.

[18] A London-born engineer by training, Bird served in the Royal Engineers until he suffered a spinal injury at Gallipoli. His first cartoon appeared in *Punch* in 1916. Fougasse, 2-3, 117-120.

[19] Morreall, 67.

[20] Karen Halttunen, "Cultural History and the Challenge of Narrativity," in *Beyond the Cultural Turn: New Directions in the Study of Society and Culture*, eds. Victoria E. Bonnell and Lynn Hunt (Berkeley and London: University of California Press, 1999), 170-171.

[21] Evelyn Cobley, *Representing War: Form and Ideology in First World War Narratives* (Toronto: University of Toronto Press, 1996), 184. Essentially this essay argues that *Punch* humorists did much to

paradoxically challenge and reinforce *both* subversion and the status quo in the targets that their cartoons attacked, hence the value of a conceptual approach that recognizes the many possible storylines yielded by pictorial humour *as well as* by the past itself.

[22] Simon Houfe, *The Dictionary of British Book Illustrators and Caricaturists, 1800-1914* (Woodbridge, Suffolk, UK: Antique Collectors' Club, 1978), 57.

[23] Houfe, 58.

[24] *Punch* Cartoon Library and Archive, "Punch Circulation Figures," London. These figures were received on 11 June 2004 via mail from the *Punch* Cartoon Library and Archive sponsored by Harrods.

[25] Holman and Kelly, 260.

[26] Pierre Purseigle, "Mirroring Societies at War: Pictorial Humour in the British and French Popular Press During the First World War," *Journal of European Studies* 31 (2001): 308-309.

[27] *Punch, or the London Charivari*, 20 January 1915 (vol. 148): 54.

[28] Houfe, 451

[29] Bevis Hillier, *Cartoons and Caricatures* (London: Studio Vista, 1970), 118.

[30] Julia Cornelissen, *The Illustrators: The British Art of Illustration, 1800-1992* (London: Chris Beetles, 1992), 70-71.

[31] Richard Holmes, *Tommy: The British Soldier on the Western Front, 1914-1918* (London: HarperCollins, 2004), xv.

[32] *Punch,* 26 January 1916 (vol. 150), 77.

[33] George Robb, *British Culture and the First World War* (London: Palgrave, 2002), 182-183.

[34] George Haven Putnam, in Bruce Bairnsfather, *The Bystander's Fragments from France* (London: Tallis House, 1914-1918), Volume 5, introduction.

[35] Bevis Hillier, 118; Jean-Yves Le Naour, "Laughter and Tears in the Great War: The Need for Laughter/The Guilt of Humour," *Journal of European Studies* 31 (2001): 268-267.

[36] Feibleman, 181.

[37] Purseigle, 296, 325.

[38] Fussell, 73.

[39] *Punch*, 30 January 1915 (vol. 148), 54.

[40] John Laffin, *Tommy Atkins: The Story of an English Soldier* (London: White Lion Publishers, 1966), xviii.

[41] Eric Bentley, "Farce," in *Comedy: Meaning and Form,* ed. Robert W. Corrigan (New York: Harper and Row, 1981), 193-194. According to Bentley, farce is notorious for its violent imagery.

[42] Gruner, 27.

[43] Lyn Macdonald, *1914-1918: Voices and Images of the Great War* (London: Penguin Books, 1988), 160.

[44] Rev. E.J. Hardy, *The British Soldier: His Courage and Humour* (London: T. Fisher Unwin, 1915), 115.

[45] Hillier, 114.

[46] John Keegan, "The Face of Battle," in *The Great War 1914-1918*, ed. Ian F.W. Beckett (London: Longman, 2001), 224.

[47] *Punch*, 20 February 1918 (vol. 154), 123.

[48] Fred W. Leigh and Kenneth Lyle, "The Army of To-Day's Alright," (Francis Day and Hunter, ca. 1915).

[49] Gary Sheffield, *Forgotten Victory, The First World War: Myths and Realities* (London: Headline Book Publishing, 2002), 247.

[50] Sheffield, 260.

[51] Leigh and Lyle, "The Army of To-Day's Alright."

[52] Sheffield, 246, 251.

[53] Macdonald, 224-225. In 1917 Bottomley stated that he had been to the trenches, which was true, and that he was a "friend of the soldier." He said that he had learnt the "truth of the trenches" and that "the war is won. Germany is beaten." Macdonald cites a sarcastic, cynical response to Bottomley's work from *The Mudlark*, the newspaper of the Royal Naval Division.

[54] *Punch*, 2 January 1918 (vol. 154), 15. The English translation reads, "I don't know what it is. It doesn't seem realistic enough."

[55] Doust, 5, 34.

[56] Winter, *Death's Men*, 103.

[57] Laffin, *The Western Front Illustrated 1914-1918* (Gloucestershire: Alan Sutton, 1991), 37.

[58] Macdonald, 201.

[59] Laffin, 37.

[60] Hillier, 118.

[61] Cobley, 18.

[62] Samuel Hynes, *A War Imagined: The First World War and English Culture* (London: The Bodley Head, 1990), 53.

[63] Holman and Kelly, 251.

[64] Willard Smith, *The Nature of Comedy* (Boston: The Gorham Press, 1930), 66-69.

[65] Leacock, 214-215.

[66] Charles L. Graves, *Mr. Punch's History of the Great War* (London: Cassell, 1919), 265.

[67] Leonard Smith, "Paul Fussell's *The Great War and Modern Memory*:

Twenty-Five Years Later," *History and Theory* 40 (2001): 249.
[68] Allyson Booth, *Postcards from the Trenches* (New York and Oxford: Oxford University Press, 1996), 83.
[69] Cobley, 15.
[70] *Punch*, 25 August 1915 (vol. 149), 167.

Appendix

British Tommy (returning to a trench in which he has lately been fighting, now temporarily occupied by the enemy). "EXCUSE ME - ANY OF YOU BLIGHTERS SEEN MY PIPE?"   **Figure 1**

THE IRREPRESSIBLES.

Tommy. "AND TO THINK THERE'S A MUSIC-HALL COMEDIAN AT HOME GETTING THREE HUNDRED QUID A WEEK FOR SINGING 'THE ARMY OF TO-DAY'S ALL RIGHT!'"

Tommy. "AND TO THINK THERE'S A MUSIC-HALL COMEDIAN AT HOME GETTING THREE HUNGRED QUID A WEEK FOR SINGING 'THE ARMY OF TO-DAY'S ALL RIGHT!'" **Figure 2**

January 2, 1918.]   PUNCH, OR THE LONDON CHARIVARI.   15

Artist (to Tommy, home on leave, acting as a model for picture to be entitled "Going over the Top"). "AH DINNA KEN WHAT IT IS. IT DOESNA SEEM REALISTIC ENOUGH. HAVE WE FORGOTTEN ANYTHING?"
Tommy. "DON'T THINK SO, GUV'NOR, ON'Y THE TOT O' RUM YER DIDN'T SERVE AHT." **Figure 3**

Budapest and the Great War: An Overview

Moni L. Riez

Abstract: Budapest in the years of World War I experienced profound economic and social deprivations. Virtually all aspects of urban life were radically transformed: health care, daily supplies, housing, employment and cultural activities. Governmental and private agencies redirected their efforts to maintaining the population under conditions of blockade and an extraordinary influx of civilian refugees and soldiers. Class inequity exacerbated frustrations but did not lead to serious social breakdown. While the emergency measures were moderately successful, the effects of the war continued for several years after the armistice.

Key Words: Budapest, World War I, British Blockade

1.   Introduction

> ...to be sure our world here is really bad. We are almost
> dying of hunger, there is nothing, we can get no meat,
> no potatoes, only corn flour, 2 kilos. The people want a
> revolution. You can't hear anything yet about peace.
> Perhaps it [the war] will never end. Things are really
> expensive... [T]he people are not able to work because
> they are hungry.[1]

In 1914 Hungary was a dynamic and populous nation. The capital, Budapest, was a growing and vibrant city, the political, intellectual and cultural centre of the nation.[2] The Hungary of 1914 was vastly larger than the one we are familiar with today; its land area was approximately two-thirds larger, with a population of 20 million citizens.[3] Budapest was home to over 930,000 residents, a thriving industry, the national parliament, and many coffee houses, theatres, cinemas and other cultural institutions. Yet all was not perfect. As in other industrial centres not all Budapest's citizens were equally well off. Under the effects of chronic unemployment and the persistent gap between wages and the rising cost of living, workers struggled to feed their families.[4] Suffrage movements on both the federal and municipal levels were under way for men and women in an attempt to establish political justice in the country and the capital. By 1917 the growth that Budapest had experienced prior to 1914 had ground to a halt and almost all sectors of its population faced deprivation. The mood since the spring of 1915 had grown glum and by 1917, as the above quotation indicates, grim conditions had been created for the inhabitants as a consequence of Hungary's participation in the Great War. In that year, as in the period prior to the war, the quality of life experienced by citizens was characterized by inequity. All citizens were still not suffering equally, though now many more of the inhabitants felt the scarcity of basic needs. This paper presents an overview of life in Budapest during the war.

2.   Infrastructure and Daily Needs

The complexity of cities is often taken for granted and overlooked. In order to run smoothly, a city must provide a complete infrastructure for its citizens, including food and fresh water, waste disposal and sewer systems, buildings, roads, lighting and institutions to preserve safety, public health and law and order. Furthermore, the people require jobs, transportation, markets for daily needs and clothing, arts and entertainment, and the institutions that provide guidance in the key times of life such as births, weddings, divorces and deaths. Even when the actual fighting is too distant to have any direct impact on the home front, wars complicate, disrupt and eventually destroy such infrastructure and institutions as the citizens increase exponentially their efforts to meet military needs.

In Budapest in 1914 the majority of citizens were very patriotic, holding parades in the city on 28 July in favour of the war and carrying signs proclaiming "Long Live the War."[5] However, these sentiments did not last long. Initially all the men between the ages of 21 and 42, except those exempted for medical and other reasons, were subject to conscription. The rapid increase in the size of the army magnified the burdens on the civilian population in both material and human terms. Thus even at the outset great sacrifice was involved as the nation mobilized to provide uniforms, food, and equipment.[6] Many of the recruits came to Budapest to receive their equipment and assignments and one could see large mobile military kitchens set up in the streets.[7] When it became clear that the men were to remain at the front for the winter, notices went out to the citizens to donate warm clothes such as toques, gloves, and stomach warmers. If citizens could not contribute in this way they were urged to offer cloth or money instead.[8] Appeals for warm clothes continued throughout the war and became more urgent as the British blockade on goods bound for the Central Powers took effect.[9] The blockade made cotton, jute and wool scarce, and this put great stress on the citizens of Budapest in trying to provide clothing not only for soldiers but also for themselves.[10] Cloth was so scarce that there was barely enough to make into sacks for transporting flour.

3.   Public Health

Resources in Budapest were also diverted to building hospitals or converting existing structures for medical purposes. The number of hospital beds in Budapest at the beginning of the war was inadequate for the rapidly growing population of the city, and it was quickly evident that more beds were needed to care for the influx of wounded soldiers. Many schools were converted into hospitals, and the new Zita Military Hospital opened its doors in 1915 with space for 3500 wounded.[11] As hospitals required sheets, food, and other supplies, appeals were redoubled for donations of both money and goods.[12] Red Cross nurse Katherine Volk arrived in Budapest in 1914 to tend to injured soldiers having travelled to Europe on an American Red Cross Mercy Ship. She described an eclectic combination of donations by the citizens:

> Word had been sent out to the public that this new hospital [the American Red Cross Hospital] had to be furnished and that supplies were needed... [M]uch of the linen that came in was lace trimmed and so frilly that it could not be used.[13]

There were appeals not only for goods to furnish hospitals but also for doctors and nurses. The increased need for care for the wounded directly conflicted with the city's own responsibility to attend to civilian medical needs, a responsibility which, as we have seen, was already stretched beyond the limit prior to the war.

One of the major concerns that arose in Budapest during the war was the threat of contagious disease. European cities had encountered several infectious disease outbreaks throughout the 1800s and were wary of the havoc they could cause in urban populations. The risk of a major outbreak was high during the war for two primary reasons. First, there was an influx of infected soldiers to Budapest for treatment. Second, the city's population increased to over one million due primarily to refugees, such as those from the Hungarian province of Transylvania who fled when the Rumanians invaded in 1916.

Measures were quickly put in place to ensure that diseases such as cholera, dysentery and typhus did not spread.[14] One measure implemented by the Hungarian Defence Minister and the Minister of the Interior was to set up 14 watch stations throughout the country where soldiers were decontaminated from lice, received a change of clothes, and were observed for any signs of contagion.[15] Soldiers were legally obliged to disembark at designated train stations where they were inspected by trained medical personnel. If soldiers chose to ignore these inspections they faced stiff fines.

In Budapest notices were posted to inform the populace of the symptoms of these contagious diseases. Other bulletins were issued to prepare healthcare workers to deal with infected patients and to prevent the spread of disease. These measures included inoculations for healthcare workers as well as admonitions by the Budapest Municipal Government as early as 10 August 1914 for health professionals to have on hand plenty of disinfectant and face masks to prevent the spread of cholera.[16] The efforts to control the spread of disease were successful throughout the war, as the incidence of infection was relatively low despite conditions rife for outbreaks.[17]

4.   Economy

Another major problem in wartime Budapest was the conflict's impact on the economy. Many industries were affected by the cessation of imports. The British blockade lasted from the beginning of the war until after the armistice in 1918.[18] Throughout that period, the British pressured neutral countries to sell their surplus to them and not to the Central Powers.[19] When it became evident that some neutrals were importing greater quantities of goods than they needed for their own domestic consumption this too was quashed. Despite the blockade, some goods entered Hungary from Holland, Switzerland, Sweden and Denmark, but they amounted to far less than the pre-war flow. For example, on 7 November 1916 the Budapest municipal council's report on the capital's food supplies spoke of the council's attempts to bring in food from the neutrals. The report mentions that in December 1915 the city had brought in 100,000 cans of cooked beef from Denmark.[20] And in June 1916, with the help of a large landowner, the city managed to obtain 397 cows to bolster its severely reduced supply of milk. The report states, however, that now they could no longer import meat or live animals from neutral countries because of the blockade and transportation difficulties. It also spoke of failed attempts to buy coffee from Sweden; Great Britain permitted the importation from neutral countries of only enough coffee to satisfy the minimum demand. After long negotiations, Budapest managed to secure a very small amount of coffee from Sweden that would be transported via Switzerland, which had given permission to transport only a minimal amount.[21]

Thus the blockade had a major impact on Hungary and Budapest. It had an immediate impact in 1914 with the cessation of most imports. By 1916 it had increased in severity, making ever more difficult any attempts to acquire resources for both daily survival - such as clothing, food and coal - and for military purposes. The blockade became complete in 1917 when the United States entered the war.

The lack of imports and a number of other factors initially created unemployment and underemployment in the city. Scarcity of raw materials for some industries, such as textiles, led to lay-offs or reduced working hours and thus lower wages.[22] One example which illustrates this problem is that of the Brassói cellulose factory. Unable to secure raw materials, by 13 September 1914 the company had to reduce its workforce from 395 to 77. This in turn affected other businesses and industries that relied on Brassói, such as small shoe factories, which were likewise forced either to close or significantly reduce the number of workers.[23] Some factories were able to convert to military production, but that also meant temporary layoffs while the factory was retooled; and in some cases when they resumed operations fewer workers were initially needed.

Another factor which led to unemployment was the induction of men for military duty. In some smaller factories, when their best skilled workers were drafted the shop had to close, causing the remaining workers to lose their jobs. This applied to retail stores, engineering firms, lawyers' offices, and other specialized services.[24] The federal and municipal governments cancelled any unnecessary building projects to save money for the costly war effort. To further reduce expenditures the government placed a wage freeze on the salaries of civil servants, a hiring freeze for most civil service positions, and mandatory reduced wages for all new hires.[25] As war casualties mounted more men were called up for military service, and as the war progressed, the age limit increased. By the end of 1915 unemployment was no longer as much of a concern since so many men were now at the front, the war industries were in full swing, and many jobs were available. Nevertheless, the issue of wages falling below the level necessary for sufficiency remained a problem throughout the war and after.

The economic impact of unemployment and underemployment in the first year of the war was particularly great for the working class in Budapest, as prices skyrocketed from the first days of the war through 1918. But even the middle-classes were affected by the wage freeze and rising prices. Price increases from 1914 to 1918 are well documented.[26] For example, from 1914 to 1918 the price of beef rose by 460 percent, potatoes by 271 percent, eggs by 1186 percent, a woman's coat by 550 percent, children's shoes by 300 percent, coal by 423 percent, matches by 471 percent and soap by 530 percent. Wages rose much more slowly, and in fact few wages rose sufficiently to cover the increase in the cost of living. A butcher's income rose 138 percent in the same four years; however, in the first two years it saw no increase. The wages of some white collar workers increased 365 percent during the war; but again this rise began only after 1915.[27] Along with the devastating price increases, chronic lack of supply meant queuing in long lines in hopes of buying basic necessities.[28]

Although the middle classes - many of whom worked in the civil service - were severely affected by the wage freezes, they still had some advantages over the working classes. Those whose residences gave access to gardens could at least raise some of their own food, such as vegetables, fruits or rabbits, while poorer classes, relegated to cramped one or two bedroom flats, could not supplement what they bought with home-grown substitutes. A case in point is Dr. Verress Endre, who accepted a job with the Budapest Archives and moved to the city in 1915. His letters to his mother, who lived in Kolozsvár, frequently referred to concerns with food scarcity and the rise in prices. However, his letters also reveal that being part of the middle class afforded him better opportunities than most people in Budapest. In October 1916 he described the house that he bought for his family in Budapest, emphasizing particularly the garden, and happily told his mother that it included several fruit trees, red currant bushes and strawberry plants. A month later, when shortages and prices were reaching or had surpassed crisis proportions for most citizens, he wrote that food was accessible but with difficulty, and he further explained that on Sundays the family ate the rabbits they had raised themselves.[29] The city government did try to assist people by selling seeds at very low prices and allowing them to garden newly designated plots of land on the city's boulevards and, eventually, the city race track.[30] The effectiveness of this effort to alleviate the food crisis is not known.

It is easy to blame the federal and municipal governments for the plight of the citizens. However, from the war's beginning the municipal government was aware that the war would make life more difficult for its citizens and it energetically searched for solutions. As soon as the war began Mayor István Bárczy announced on 31 July 1914 that the city was setting up the Városi Népsegittö Irodá (The People's Municipal Social Service Office), which was to open on Monday 3 August 1914 to help families affected by the war. He also asked municipal workers to donate one percent of their salaries to help the families of colleagues serving in the military. Other organisations, such as the teachers and doctors associations, offered similar assistance. Charity concerts were held to benefit the poor, along with drives for donations of clothing and food.[31] Wealthy munitions manufacturer Manfred Weisz established a kitchen to serve the unemployed and families affected by the war; from the beginning of the conflict it provided 1000 meals daily. He fulfilled his pledged to continue this effort throughout the war, and in fact he expanded the service to 1500 meals a day. Ironically, with the profits he made in war manufacturing he was well able to afford the expense. Meals at these war kitchens were either free or at a much reduced cost. The city at one point even set up mobile kitchens to help women who were ill or working to provide their families at least one hot meal a day.[32]

The city administration also persistently lobbied the federal government for measures to deal with the growing increase in the prices of goods. On 25 June 1915 the mayors of the various Hungarian cities gathered to discuss their major concerns. While the federal government had reluctantly put in place maximum prices on some goods such as flour, by 1 December 1914 the mayors all agreed that even these prices needed to be lowered.[33] Their conclusion was not that the policy was deficient but that the government set the maximum too high. Another primary concern of the mayors - who were acutely aware of the consequences of the grain shortage during the winter of 1914 - was to ensure that there was enough grain for their citizens for the upcoming year. Their solution was to lower the maximum price of grains from 40.50 korona to 36.60 korona and then further reduce it to between 34 and 30 korona per quintal.[34] The rationale for such a move was that the wages for most of their citizens were not rising fast enough to accommodate the rapid price increase of grains, which were a staple of many family diets. However, this reasoning was overly simplistic. Prices were increasing because the costs of producing, transporting, milling and selling the finished product had all increased as well. More grain would not become available through price reduction; on the contrary, when prices were reduced the cost of production was not being covered, and production fell. One solution applied in Great Britain was for the government to provide subsidies for the producers; in Hungary, however, the government's meagre financial resources and lack of access to loans made this approach impossible.[35]

Persistent class differences and the inequality of suffering caused further frustration.[36] The wealthy, such as financiers, manufacturers and landowners, were able to make profits from the war, buy goods on the black market, and grow food on their property or even hunt game on their estates. This point is well illustrated in a letter dated 14 September 1918 written by composer and folklorist Béla Bartók to a friend:

> Dear Professor,
> After a week at home, I went to spend a fortnight in lordly splendour with the Kohners. Three footmen and a parlour maid served at dinner, and there were two menservants and a chambermaid to tidy my

room. Coaches, horses, food, baths, cigarettes, wine, real
coffee - plenty of everything of the best.

> And these people have such a gift for enjoying
> their affluence that one almost forgets to be angry at the
> unequal distribution of wealth.... They had 4 women to
> do the laundry.... And this tremendous display of luxury
> and labour is all for the benefit of the Kohner couple and
> their 3 children! N.B. They have a town house in
> Budapest as well, and another house on their estate in
> Hont County.
> The Baron went out hunting each morning and
> came home every day with a bag of 25 to 30
> partridges.[37]

The lifestyle enjoyed by the Baron and his family was vastly
different from that of the majority of the people of Budapest. The
government's monthly morale reports of 1917 and 1918 provide detailed
information from letters read by government censors as to public opinion
throughout Hungary. The reports examined the views of the people both in
the city and rural areas and also the intelligentsia, business class, and
poorer segments of society. Attached to the reports were quotes from
letters, original censored letters, and letters that had secret messages
written in them using number combinations, lemon juice, or secret
compartments.[38] It is clear that either the wealthy knew better than to
complain because of the censorship or that they, like the Kohners, had
little to complain about. By contrast, the urban intelligentsia and the lower
classes by 1917 were calling constantly for peace, their lives deranged by
high prices and scarcity. The quote from the June 1917 letter at the
beginning of this paper is typical of comments about life in Budapest at
this time. It is fair to assume that many of these people knew censorship
was in force. In fact their complaints about a lack of government financial
help were routinely passed on by the censor to appropriate government
officials who, if they felt a complaint was baseless, would send a letter
rebuking the complainant. By this point, however, many appeared to be
beyond caring about the censors. People were deeply frustrated, angry and
threatened by the food scarcity.[39]

While the war curtailed the ability of citizens to procure daily
necessities, it did not put an end to cultural life in Budapest. Many forms
of entertainment continued and were well attended. For people who were
less well-to-do there were movies and the zoo, and one could still buy
books and periodicals ranging from daily newspapers to literary journals.
Yet the war affected these resources too, as newspapers had to reduce the

number of pages they sent to print because of a lack of paper. Theatres continued to stage new plays, and in 1916 according to a police report they were well attended; one exception was when all forms of entertainment were shut down for a week following Franz Joseph's death.[40] Concerts were held, and Béla Bartók wrote and endured the frustrations of putting on his operetta *Bluebeard's Castle* in 1917. Since there was a shortage of materials to produce lighting for the city, street lights were scaled back as were the hours of operation for restaurants, coffee houses, dance halls, and cabarets. In fact by 1916 all establishments had to be closed by 1 am and they could not open before 5 am. These restrictions were not trivial. Prior to the war the literati and academics of Budapest centred their socializing on the approximately 600 coffee establishments in the city, many of which were open twenty-four hours a day, 365 days a year.[41] Thus the cultural life of Budapest was severely curtailed. Novelist Margit Kaffka summarized the situation in a letter written to one of her literary friends in Budapest in October 1916. "I would like to be in Pest. Where do you meet now in coffeeless coffeehouses; where should I look for you?[42]

Although Budapest could not display a sign on the door saying "Closed for Business until War's End" - for too many people lived there and it was the capital, the heart and soul of Hungary - yet the war dramatically changed the lives of the citizens. Wartime services became the focus of the city's life: ensuring that soldiers did not spread contagious diseases from the warfront to the home front, helping the families who were affected, and producing military supplies and equipment. For the middle and working classes the war increased unemployment, decreased wages, and dramatically increased the price of all goods. Both the federal and municipal governments were aware of the problems faced by the people and sought solutions, but with little success. One of the main reasons for this failure was that the British blockade prevented the city from receiving the goods it needed. Nevertheless, even in these hard times people continued to seek venues to be entertained and to forget, if even for a little while, the stark reality of their day-to-day lives.

Notes

[1] Hadtörténeti Levéltár (HL) [Hungarian Military Archives], Első világháboru gyűjtemény [First World War Collection], Zensurkommission für Korrespondenzen an Kriegsgefangene in Budapest, 4447 d.
[2] When Hapsburg political fortunes were rocked by their loss in 1866 to Prussia, Franz Joseph became more willing to look for support within his

empire and ultimately signed the 1867 Compromise with Hungary, transposing the Hapsburg Empire into the Dual Monarchy. Hungary became independent in all its domestic affairs from that point on. The 1867 Compromise allowed the towns of Buda, Óbuda, and Pest to reconsider their future and by 1872 they had signed the agreement that would by 1873 see them amalgamated into one rapidly growing city, Budapest.

[3] Hungary was reduced to its current size in 1920 with the signing of the Treaty of Trianon, Hungary's peace treaty at the end of World War I. Under the terms of the treaty Hungary lost 70 percent of its land area and 60 percent of its population.

[4] The women's newspaper *Nőmunkás* (Woman Worker) issued a supplement on 15 June 1913 that examined in detail the problems faced by the working classes, especially women and children, and advocated solutions. See also I.T. Berend and Gy. Ránki, *The Development of the Manufacturing Industry in Hungary (1900-1944)* (Budapest: Akadémiai Kiadó, 1960), 39-44. Berend and Ránki go into detail regarding the problems faced by the Hungarian working classes, such as an average 40 percent increase in food prices from 1901 to 1913, poor working conditions, long hours, and only a nominal rise in salaries.

[5] Deputy Mayor Tivador Body gave a patriotic speech on the steps of city hall on 28 July 1914. In *Források Budapest Multlyából* [Primary Documents in Budapest's History], ed. H. Kohut Maria (Budapest: Budapest Főváros Levéltár, 1971), 292. At the Hadtörténelmi Muzeum Phototár's World War I Collection (Hungarian Military Museum Photograph Archives), several pictures can be found of the outbreak of the war in Budapest with parades of people in support of the war.

[6] As the losses quickly mounted this was changed by 1916 to all men between the ages of 18 and 50.

[7] Men who were called up came to Budapest from all over the country to receive their uniforms and equipment. The rail system was set up in Hungary such that Budapest was at the centre of all lines. In the Hadtörténelmi Muzeum Phototár's World War I Collection there is photographic documentation of conscripts from throughout Hungary reporting to their units in Budapest. This collection also includes photographs of clothes distribution and military kitchens in the streets of the capital.

[8] Széchényi Könyvtár [Hungarian National Library], Plakát Tár [Circular/Pamphlet Collection], World War I Collection 1914. Hereafter this collection will be referred to as PT and the year. One of the pamphlets appeals to women and girls to knit such items as snow toques and wrist

and stomach warmers, and includes patterns and step-by-step instructions. Kohut, 321.

[9] PT, 1915-1918. There are several works on the blockade and how and why it changed over the course of the war. For the official history of the blockade see Archibald Bell, *The Blockade of the Central Empires, 1914-1918* (London: HMSO, 1961); Jonathan Clay Randel, "Information for Economic Warfare: British Intelligence and the Blockade, 1914-1918" (PhD diss., University of North Carolina at Chapel Hill, 1993); W. Arnold-Forster, *The Blockade 1914-1919: Before the Armistice and After* (Oxford: Clarendon, 1939).

[10] W.G. Max Müller, "Memorandum, 30 November 1914," in *British Documents on Foreign Affairs: Reports and Papers from the Foreign Office Confidential Print*, eds. Kenneth Bourne and D. Cameron Watt, Part II, Series H, Volume 9 (New York: University Publications of America, 1989), 43. Müller was the British Consul-General in Budapest prior to the war. During the war he wrote monthly reports on daily life in Austria-Hungary. By comparing his reports to Hungarian primary sources I have found his observations and information to be fairly accurate.

[11] Kohut, 322.; PT, 1914.

[12] PT, 1914.

[13] Katherine Volk, *Buddies in Budapest* (Los Angeles: Kellaway-Ide, 1936), 96.

[14] Ibid. One of the earliest public notices issued by the Hungarian Minister of the Interior, János Sándor, was on 30 July 1914 to city officials reminding them of what rules needed to be followed with regard to contagious diseases.

[15] József Mailáth, *A Harctéri Betegmegfigyelő Állomásokról* [The Battlefront Watch Stations] (Sátoraljaújhely: Landesmann, 1915); Gyula Tarnay, *A Miskolci Megfigyelő Állomás* [History of the Miskolc Watch Station] (Miskolc: Klein and Ludvig, 1916).

[16] PT, 1914.

[17] Gustav Thirring, *Budapest Székesfőváros Statisztikai Évkönyve 1913-1920* [City of Budapest Statistical Yearbook 1913-1920] (Budapest: Budapest Statistical Bureau, 1923), 80-81.

[18] For further information on the continuance of this economic strategy, see Charles Paul Vincent, "The Post-World War I Blockade of Germany: An Aspect in the Tragedy of a Nation" (PhD diss., University of Colorado, 1980), and Clifford R. Lovin, "Food, Austria, and the Supreme Economic Council, 1919," *East European Quarterly* 12 (1980): 475-487.

[19] For a detailed look at the neutral countries and the blockade see Edgar Turlington, *Neutrality: Its History, Economics and Law - The World War*

Period (New York: Octagon, 1935).

[20] Kohut, 342.

[21] Ibid., 342-345.

[22] Sándor Gábor, "Adatok a Dolgozók Helyzetéről és Megmozdulásairól az Első Világháború Kezdetén" [Information on the Workers' Situation from the Beginning of World War I], *Partörténeti Közlemények* 21 (1974): 159-88; Iren Nevelő, "A Munkásosztály Helyzete Magyarországon az Első Világháború Idején" [The Working Classes' Situation in Hungary During World War I], *Századok* 99 (1965): 138-153.

[23] Gábor, 163.

[24] Ibid., 161-162.

[25] Emma Iványi, *Magyar Minisztertanácsi Jegyzőkönyvek az Első Világháború Korából* [The Hungarian Cabinet Ministers Meetings' Minutes from the Period of World War I] (Budapest: Akadémiai Kiadó, 1960), 74-75.

[26] The World War I PT or Circular Collection has many circulars listing maximum prices as well as prices the government would pay for metal goods offered for the war effort.

[27] Benő Gál, *Az Árok Alakulása az Első Világ Háboruba* [The Development of Prices in the First World War] (Budapest: Schimkó, 1925), 2. This booklet, published shortly after the war, differs from most books of the period because it not only states the prices of goods and price increases during the war but also weekly wages; moreover, in the final section it compares the two.

[28] While Hungary was an agrarian country and agricultural products were its main export, major crop failures occurred throughout the war period starting in 1914 when there was serious damage due to poor weather. Moreover, Hungary needed to export some of its agricultural goods in order to obtain necessary supplies such as coal from Austria or Germany.

[29] OL Private Letter Collection P. 1569 2d. 6 tetel. Almost all of his letters refer to some aspect of food, such as advising his mother on what to do with the extra food she had grown.

[30] PT, 1915-1916.

[31] Ibid., 1914-1918.

[32] Kohut, 337-338.

[33] Hungarian Prime Minister István Tisza had received requests to establish maximum pricing from the very beginning of the war, but he demurred. He explained his rationale in a 1 August 1914 letter to a government official in the city of Pozsony: "Dear Friend! The city officials need appropriately deal with the problem of the increasing price of food.... [T]here is the possibility of the government fixing the prices,

but this is a double-edged sword, because as soon as a maximum price is fixed it will instantly become the minimum price." István Tisza, *Gróf Tisza István Összes Munkái* [The Collected Works of Count István Tisza], Series 4, Volume II, 37-38 (Budapest: Franklin, 1924).

[34] A standard measure at that time in Hungary, a quintal weighed approximately 100 kilograms. Papers of the Prime Minister, Hungarian National Archives, K26 cs1030, 3550 res. 1915.

[35] For further analysis of the food crisis and the method Britain employed to find solutions see Margaret Barnett, *British Food Policy During the First World War* (Boston: George Allen & Unwin, 1985); P.E. Dewey, *British Agriculture in the First World War* (London: Routledge, 1989); E.H. Whetham, *The Agrarian History of England and Wales, 1914-39* (Cambridge: Cambridge University Press, 1977), chaps. 5-8.

[36] Nándor Kozma, *Hadimilliomosok* (Budapest: Muskát Béla Könyvnyomda, 1918). The book scathingly points this out and lists the various war millionaires and how they made their money. Manfréd Weisz, who was involved in industries and providing military goods, is among the most prominent of the war profiteers cited in this work.

[37] Béla Bartók, *Béla Bartók Letters*, eds. János Deményi et al. (London: Faber, 1971), 139-141.

[38] HL, 4447 d. and 4448 d.

[39] HL, 4411-4414 d. These files contain government letters responding to requests for financial aid.

[40] Kohut, 334-337.

[41] John Lukacs, *Budapest 1900: A Historical Portrait of a City and Its Culture* (New York: Grove Weidenfeld, 1988), 148-153. For further information on early twentieth-century Budapest and its culture see also the following: Judit Frigyesi, *Béla Bartók and Turn-of-the-Century Budapest* (Berkeley: University of California Press, 1997); and Mary Gluck, *Georg Lukács and His Generation, 1900-1918* (Cambridge, MA: Harvard University Press, 1985).

[42] Petőfi Irodalmi Museum [Hungarian Literary Museum], Margit Kaffka Letters, V3181/71/4.2.

Part II

Victims

War Survivors' Fractured Identities in
Hiroshima mon amour

Brigitte Le Juez

Abstract: War survivors can suffer from a certain loss of memory and/or an inability to articulate their experience. These symptoms of trauma can be either permanent or temporary. The personal experience of the survivor, set in a particular cultural context, can also lead to a crisis in the individual's sense of his/her national identity, by which I mean identity defined in terms of the country to which one belongs. Many artists have tackled the subject, more often than not from an autobiographical perspective. This essay examines how this question has been represented in literature and cinema, namely through one very notable *ciné-roman*, *Hiroshima mon amour* (1958).[1] The film centres on an encounter in Hiroshima between a French woman and a Japanese man, which triggers memories of events during the Second World War in France. This article examines the role of testimony in this dual circumstance, as well as the literary and visual devices used in the film to represent the different qualities and effects of testimony, especially in terms of personal trauma and loss of identity. Also considered are the historical context of this *ciné-roman* and its wider significance with regard to the political implications of cultural trauma on an individual's sense of national identity.[2]

Keywords: Testimony, War Memories, Identity, Alain Resnais, Marguerite Duras, Victimization

1.   The Role of Testimony

The perspective of *Hiroshima mon amour* is that of a personal experience, and specifically the testimony of a survivor of the Second World War. Before making *Hiroshima mon amour*, Alain Resnais had already directed a film about the Second World War, his famous 1955 documentary on concentration camps, *Nuit et Brouillard (Night and Fog)*. In *Night and Fog* he had managed, through a powerful effective contrapuntal juxtaposition of peaceful colour images of Auschwitz ten years after the war and unbearably precise black and white archive pictures, not only to refuse the effacement of memory, but also, in a manner of speaking, to summon the dead from their graves to testify. In this documentary, the anonymous faces of the inmates, reduced to a common condition and expression of suffering, represent the crushing of identities, both individual and national, in the concentration camps. The voice-over, a text written by a survivor of the camps, the poet Jean Cayrol, and delivered in a monotone, as if it came from the collective grave of the victims, expresses the permanence of their common trauma.

When Resnais was commissioned to make a documentary on the atomic bomb by the producers of Argos Films, for whom he had made *Night and Fog*, he had difficulty at first in imagining a new way to deal with the subject of war that would do it justice. He therefore decided, as he had done in the past (and would continue to do), to ask an established literary figure to help him with the script of what was to be his first feature film. He chose Marguerite Duras, who, in her own writing, had shown a clear interest in using personal experience to explore the human psyche. One of Duras's literary techniques consists of the insertion of autobiographical material into her fiction, hence her interest in the use of testimony. She wrote in the synopsis of the screenplay for the film that the characters' "personal story, no matter how brief it may be, must always dominate HIROSHIMA."[3] The love scenes at the beginning of the film are interspersed with images of Hiroshima after the impact of the bomb, with the physical evidence of the devastation either evoked or plainly visible on human bodies. The visual effect is similar to that of the alternation of archive pictures with colour images in *Night and Fog*: it gives the viewers the uncomfortable feeling that the most horrific events of war can be forgotten, that life goes on, as the natural growth of plants testifies (a theme found in both *Night and Fog* and *Hiroshima mon amour*). The personal story, therefore, is of crucial importance, because, drawing on an intensely emotional experience, it is presented as a counter to nature's indifference, as a trauma which, precisely, cannot be erased.

Hiroshima mon amour examines the short but intense and compelling encounter between a Japanese man and a French woman who, through the emotions triggered by this encounter, recalls a personal trauma suffered during World War II in the French town of Nevers. Her trauma is directly related to the attitude of her compatriots at the time of the liberation of France, and to the death of her lover, a German soldier killed by an unidentified, but undoubtedly French, sniper.[4] The lovers had planned to go to Germany to get married after the war; instead, she found herself unable to mourn him properly, imprisoned as she was in a general atmosphere of hostility and violence. Like other women who had had relationships with German soldiers, her head was publicly shorn in the town square, and her dishonoured family could think of nothing better to do with her than to lock her up in their cellar, pretend she was dead and, after her hair had grown again to a so-called "decent" length, let her disappear from their self-righteous and typically provincial French lives. The film joins her twelve years later, in Hiroshima, where she is acting in a film.[5] It is her last day in Hiroshima. The night before, she has met a

Japanese architect. The first verbal exchange we witness between them is on the subject of Hiroshima, but the dialogue is about to shift to her recounting, for the first time ever, her ordeal twelve years earlier in France.

I have focused thus far on the woman's story because it largely dominates the film; but this does not mean that the story of the Japanese man is not relevant. He was a soldier away from Japan at the time of the bombing, and was spared. His entire family, however, was killed. He is, therefore, interested in *listening* to the testimony of a civilian war-crime witness, while she needs to *give* her testimony. Viewers are sometimes frustrated that his experience is left untold, but in reality his presence in the film is crucial as he acts as a catalyst for her testimony. During her narration he occasionally acts the part of the German soldier. Spurring her on to speak, he asks at one point: "When you are in the cellar, am I dead?"[6] In this way he shares her burden and her solitude. Their common pain at the war crimes they either witnessed or of which they were, directly or indirectly, the victims, is what brings them close together, possibly unconsciously. For the first time since her ordeal, the female protagonist is able to communicate what happened to her in Nevers; she has never been able to speak about it, even to her husband. The unexpected encounter with the man in Hiroshima revives the memories of the events of Nevers. And her strong attraction towards him encourages the possibility of remembrance: he inspires in her the same combination of passion and pain that she once felt for the German soldier. She suddenly has a sense of having found, for the first time since her trauma occurred, a responsive interlocutor: "Naturally you can also understand this," she says to him, referring to the near-madness of the grief and resentment she felt following the events in Nevers.[7] These have never left her. She tells her Japanese lover that Nevers is "the thing" she thinks about least and dreams about most.[8] Her story becomes the articulation of what has been, until now, inexpressible.

The woman thinks she understands the pain and horror of Hiroshima because she believes that Hiroshima is linked to her personal history. But in the early stage of their relationship, when she speaks about Hiroshima, the man continually denies the knowledge and understanding she claims to have of Hiroshima. For Duras, it is "impossible to speak about HIROSHIMA. One can only speak about the impossibility of speaking about HIROSHIMA. *The knowledge of HIROSHIMA* being, at the outset, presented as an exemplary delusion of the mind"[9] - hence the choric line in the dialogue between the two protagonists: "You saw *nothing* in Hiroshima. Nothing. / I saw *everything*. *Everything*."[10] He will only begin to understand why she may legitimately feel this way when she starts relating her story to him. According to Shoshana Felman:

> As a relation to events, testimony seems to be composed of bits and pieces of a memory that has been overwhelmed by occurrences that have not settled into understanding or remembrance, acts that cannot be constructed as knowledge nor assimilated into full cognition, events in excess of our frames of reference.[11]

The French woman thus concentrates her attention on Hiroshima because it is both related to her experience and removed from it; she feels herself to be sufficiently distant to discuss Hiroshima objectively, something she cannot do with her past experience in Nevers. At first, instead of talking about herself, she describes the plight of the victims of Hiroshima, using secondary sources of information, like the museums in Hiroshima or the news-reel images of the victims of Hiroshima, punctuating her various examples with empathic phrases, such as, "I did *see them,*" or, "I did not make *anything* up," or, "I know *everything*" - claims which are all in turn denied by her partner: "You did not see anything." "You made everything up." "You know nothing." [12]

2.   Personal Trauma and Identity

When she eventually offers her own narrative, the French woman does not give practical, objective details such as dates and figures. She no longer offers borrowed memory but rather a recollection of an agonising moment in her life. She describes how she stayed with her German lover until his body went cold, and how she went mad with grief after his death. She describes what was done to her by the people of Nevers, alternating direct and indirect speech, a device denoting the mental numbness generated by the traumatic pain she suffered: "They shear my hair carefully, all of it. They think it is their duty.... Someone says she should be paraded through the town."[13] She seems to have been more or less oblivious to the injustice done to her then, because her thoughts were focused principally on the idea of the death of her lover. It is clear that because she repressed the trauma she did not undergo a natural mourning process. The tone of her voice while telling her story is also reminiscent of that of the commentary of *Night and Fog*, reminiscent, therefore, of that film's "voice of the dead." Looking at her own reflection in the bathroom mirror, she says to her dead lover: "You were not quite dead. I told our story.... It *could* be told, you see."[14]

The device of the mirror is important. The mirror, like the river which plays a big part in *Hiroshima mon amour*, has been used elsewhere in cinema (as in literature) to signify death, or the passage from the world of the living to the world of the dead. In Cocteau's *Orphée* (1949), for instance, when the hero tries to reach his deceased wife, Eurydice, he passes from one dimension to the other through his bedroom mirror. The orphic theme is also one that was central to a film which came out at the same time as *Hiroshima mon amour* (in 1959) and which received the Palme d'Or in Cannes that year, *Orfeu Negro*, by Marcel Camus. And Claude Lanzmann's *Shoah* (1984), which deals with the testimony of the victims of the Holocaust, also resembles Resnais's work in this regard, as Emma Wilson points out:

> Lanzmann films the verdant countryside and the extraordinary peace which surrounds the place of horror. We see a middle-aged man in a boat which is steered by an elderly boatman along the course of the dark river Ner. The image signifies a crossing over. The river recalls for a moment the river Styx which mortals cross to reach the Underworld in Greek mythology.... Lanzmann illustrates visually the path we will take to enter his film and the Underworld which will be his subject.[15]

Jean Cayrol, the script-writer of *Night and Fog*, was, in his literary work, interested in the figure of Lazarus, another mythic figure associated with death and resurrection. The cathedral mentioned by Duras in the screenplay was originally supposed to be that of Saint-Lazarus.[16] This symbolism is related to an aspect of testimony, that of the sense of guilt which haunts survivors, and which, according to psychiatrists in this field of study, is quite typical amongst people who have witnessed torture or murder during war but who, having survived, feel as if they have betrayed their companions, as if they have left them behind. The French word *"hantée"* (haunted) is emphasised in the indications given by Duras during the crucial narration scene set in a café. The French woman considers herself a traitor to her German lover; it is as if telling their story had undermined the intensity of what happened between them, both in terms of extreme passion and extreme pain, reducing and making banal her trauma in the process. Walking alone in the dark in Hiroshima, she recalls scenes of her first love and castigates herself: "Sad little novelette, I consign you to oblivion.... Little girl from Nevers. Little slut from Nevers.... Shorn little lamb of Nevers I consign you to oblivion tonight."[17]

According to Willy Szafran, the transformation of an endless mourning - by which he means a mourning that knows no end because of a sense of guilt which is never overcome - into a "normal clinical" mourning, could be achieved through a phenomenon referred to as "*historisation*," in which the survivor attempts to transcend his/her personal drama by placing it in its historical context.[18] This is what the French woman unwittingly endeavours to do: by focusing her attention on and speaking of the horror of an historical event, she slowly finds herself able to speak about her own trauma and then able to deal with it - which helps her, eventually, to face her loss. The end of her act of witnessing may be seen as the end of her mourning, if not altogether as her recovery. This process is represented visually by Resnais in a succession of images and sounds mixing Hiroshima and Nevers. As the woman walks the streets of Hiroshima alone in the darkness all that is heard is her interior monologue, addressed to both her German lover and her Japanese lover (who is following her from a distance). This monologue is accompanied by a succession of images of Nevers and Hiroshima, underlining her experience of despair in Hiroshima, which really started many years previously in Nevers. The soundtrack (Japanese music, street noises and outside voices), accompanied and juxtaposed with some easily identifiable French cultural features (such as street names, plane-trees, slate-roofed houses), emphasises the impossibility of forgetting, or of recovering from, the trauma of war. The realisation and acceptance of this eventually allow the protagonists to recognise their identities, including their national identity.

There is a double irony in the temporal indications concerning the woman's grief: it spans the time of the liberation of France, whilst her own liberation, represented by her departure from Nevers and her arrival in Paris, coincides with the bombing and the *news* of the bombing of Hiroshima. It is precisely because Hiroshima for her is synonymous with extreme pain, the kind she suffered to the point of near-madness whilst other people were celebrating the liberation of France, that she seems to have detached herself totally from her French identity, especially from the *renewed* sense of French identity generated by *la Libération*. Describing the evening of the day of the cropping of her hair, she says: "*La Marseillaise* is being sung all over the town. Darkness falls.... My father's pharmacy is closed because of the disgrace. I am alone. Some people are laughing. In the night I make my way home."[19] She clearly expresses here the total break between herself and her compatriots, the avenging crowd responsible for her humiliation, and also the distance between herself and the geographical and political contexts of her situation. Without emotion, she describes what is going on and what she is doing, cut off from her people, place and time.

Although the nationalities of the characters are underlined throughout the film, the notion of a national identity is clearly questioned here. The French woman has become rootless, and identifies with all the victims of war, without any distinction of nationality. The film she is acting in - the reason for her presence in Hiroshima - is about peace. She tells the Japanese man that it is not a French film but an "international" one.[20] While she is constantly referred to as "la Française" in the script, she is not really seen as a representative of her country. She plays the part of a nurse in the movie being filmed, a role she already had in Nevers and which brought about her encounter with the German soldier (he had hurt his hand and had gone to her father's pharmacy to have it looked after). This time, however, she is a Red Cross nurse, a role which develops the character's statelessness (Duras adds in the script commentary that she should be perceived as the eternal nurse of an eternal war).[21]

This loss of identity is further reflected in the fact that none of the characters has a name. She is "Elle" (she) and he is "Lui" (he). She mentions that, after her lover's death, she kept repeating his German name, but the name is never given.[22] Their sense of individual identity has been crushed by the war. Yet, as the testimony ends, a new identity emerges. Indeed, the film concludes with the two protagonists about to separate and giving each other a name for the first time: "Hi-ro-shi-ma. Your name is Hi-ro-shi-ma," she says to him. "Yes, it's my name," he replies and adds: "And your name is Nevers, Ne-vers-en-France."[23] The sites of their respective traumas have become their individual identities. This is how it must be: there is no possibility of forgetting the trauma, because that place is, as the Japanese man puts it, the place where one became the person one is and will continue to be.[24]

3.   Political Implications

Duras and Resnais complement each other perfectly in this film. Duras deals with the causes of trauma in times of war, such as xenophobia, murder, and specific brutalities like that of the denunciation and punishment of women at the end of the war for their relationships with German men. Indeed, the choice of characters here is significant, as both the French woman and the Japanese man would hardly have been perceived as war victims at the time the film was made - rather as collaborator and aggressor, respectively. In an interview, Resnais spoke of both his and Duras's deliberate attempts at creating anti-heroes.[25]

Resnais, through his unique filming technique, establishes a parallel between Hiroshima and Nevers, thus conferring a unity upon completely different places and situating trauma within a cultural context. Duras and Resnais fuse together two times, past and present; two places, France and Japan; and two stories, thus representing the universality and the timelessness of trauma and crises in identity for war survivors. Duras also explores the notion of "the enemy," specifically in the scene where the protagonist explains why her compatriots shaved her head at *la Libération*: "My dead love is an enemy of France."[26] It is significant that she does not say "my dead lover." What is being punished, therefore, is what she allowed herself to *feel*. This injustice is what made Duras consider having the notion of patriotism clearly denied in one of the woman's lines: "I wish I no longer had a homeland" - a line which was not retained in the final script.[27] Her rejection of France is rendered instead in the evocation of the joyous celebrations taking place while she is locked in the cellar. She recounts the cacophony of the playing of *La Marseillaise* and the ringing of the church bells. She also describes how society passes by above her head; imprisoned in her cellar, she is profoundly the outsider.[28] In the woman's narrative the heart of Nevers, "la Place de la République," is set in stark contrast with the heart of Hiroshima, "la Place de la Paix," as if she were unable to reconcile the notions of a French republic and peace.

One should not forget that at the time the film was made (1958), France, as both victim and perpetrator of war atrocities, was far from having dealt with all its World War II demons, and was indeed in the process of dealing with new ones, those generated by the war of independence in Algeria (which lasted from 1954 to 1962). It is worth noting that the Battle of Algiers took place in 1957, the year given by Duras for the story, and that Duras joined the *Comité des intellectuels contre la poursuite de la guerre d'Algérie*, an anti-war committee whose rallying call was precisely the refusal to condone the oppression of the Algerian population carried out in the name of France. One need also take into account that the war in Indochina, which ended in 1954, only four years before the film was made, was a conflict to which Duras would have been particularly sensitive, since she was born and had grown up in Indochina.[29]

These elements reinforce the notion of autobiography and, more importantly, of testimony which, as I mentioned earlier, was important to Resnais. In 1966, Resnais asked Jorge Semprun, who not only had been a prisoner of the concentration camps but also an anti-Franco activist, to write the script of his next film, *La Guerre est finie* (*The War is Over*). In *L'Ecriture ou la vie,* Semprun revealed that his script for Resnais was strongly autobiographical, taking the form of a personal testimony on trauma and identity.[30] Between 1955 and 1966 the representation of war survivors' fractured identities was the central subject of Resnais's films, as expressed through the testimonies of both his characters and his script-writers - a fracture which was the result of the difficulty in reconciling one's sense of a national identity with cultural trauma in times of war.[31] Far from offering any totalising explanations, Resnais was insisting that these questions ought to remain open to further reflection. By enabling us to sympathise with one exemplary experience of cultural trauma, he addressed difficult social and historical issues, and at the same time encouraged new ways of considering our moral responsibility towards recognising and helping the sufferers of such trauma.

Notes

[1] The term *ciné-roman* (which can be translated as filmic novel) was created by the French novelist Alain Robbe-Grillet when he himself started his cinematic work with Alain Resnais, several years after Marguerite Duras. Resnais therefore can be seen as the inspiration for this original conception of cinema, and the term can be applied to his work with Duras. The term *ciné-roman* is now used to describe a novel written to be filmed, with filmic images in mind, but distinct from a scenario insofar as it does not contain images or technical details. The *ciné-roman* is an independent text which can be read as a novel on its own, but which is inseparable, from the point of view of its reputation, from the film. For further considerations on this subject see Brigitte Le Juez, "Filming with Writers: Alain Resnais's Literary Cinema," in *Reading Across the Lines*, eds. C. Shorley and M. McCusker (Dublin: The Royal Irish Academy, 2000), 221-228.
[2] On the subject of cultural trauma, see Jeffrey C. Alexander et al., *Cultural Trauma and Collective Identity* (Berkeley: University of California Press, 2004).
[3] "Toujours leur histoire personnelle, aussi courte soit-elle, l'emportera sur HIROSHIMA." Marguerite Duras, *Hiroshima mon amour* (Paris: Gallimard, Folio, 1960), 12. This is my translation. The English translation of dialogues is taken from the subtitles offered in the English version of the film (Argos Films).
[4] Duras gives the dates of July 1944 and of 6 August 1945. Ibid., 27, 132.
[5] The bombing of Hiroshima happened in August 1945 and the time given by Duras for the story is August 1957. Ibid., 9.
[6] "Quand tu es dans la cave, je suis mort?" Ibid., 87.
[7] "C'est vrai que ça aussi tu dois le comprendre." Ibid., 59.

---

[8] "Nevers, tu vois, c'est la ville du monde, et même c'est la chose du monde à laquelle, la nuit, je rêve le plus. En même temps que c'est la chose du monde à laquelle je pense le moins." Ibid., 58.

[9] "Impossible de parler de HIROSHIMA. Tout ce qu'on peut faire c'est de parler de l'impossibilité de parler de HIROSHIMA. *La connaissance de Hiroshima* étant a priori posée comme un leurre exemplaire de l'esprit." Ibid., 10.

[10] "Tu n'as *rien* vu à Hiroshima. Rien. / J'ai *tout* vu. *Tout*." Ibid., 22.

[11] Shoshana Felman and Dori Laub, *Testimony: Crises of Witnessing in Literature, Psychoanalysis and History* (New York: Routledge, 1992), 5.

[12] "Je les *ai vues,*" "Je n'ai *rien* inventé," "Je sais *tout*." Duras, 27-30.

[13] "Ils me tondent avec soin jusqu'au bout. Ils croient de leur devoir de bien tondre les femmes.... Quelqu'un dit qu'il faut la faire se promener en ville." Ibid., 96-97.

[14] "Tu n'étais pas tout à fait mort. J'ai raconté notre histoire.... Elle était, vois-tu, racontable." Ibid., 110.

[15] Emma Wilson, *French Cinema since 1950: Personal Histories* (London: Duckworth, 1999), 86.

[16] The figure of Lazarus is twofold. In John 11-12 he appears as the man whom Jesus restored to life. In Luke 16: 12-31 he is the sickly beggar who lay at the rich man's gate and survived on the scraps from his table until he died and was raised to heaven; of him the rich man, eventually finding himself in hell, pleaded to return to the world of the living to warn his rich brothers against their fate. The cathedral in the final version is that of Saint Etienne (Saint Stephen), considered to be the first Christian martyr.

[17] "Histoire de quatre sous, je te donne à l'oubli.... Petite fille de Nevers. Petite coureuse de Nevers.... Petite tondue de Nevers je te donne à l'oubli ce soir." Duras, 118.

[18] Willy Szafran, "Les morts dans les témoignages de la vie concentrationnaire (*Les Dibboukim*)," in *Écriture de soi et trauma*, ed. Jean-François Chiantaretto (Paris: Anthropos, 1998), 141: "sur le plan clinique je m'étais formulé une hypothèse selon laquelle le deuil chez les rescapés serait possible grâce au phénomène de 'l'historisation' par lequel le rescapé tenterait de transcender son drame personnel en plaçant ce dernier dans son contexte historique."

[19] "On chante *La Marseillaise* dans toute la ville. Le jour tombe.... La pharmacie de mon père est fermée pour cause de déshonneur. Je suis seule. Il y en a qui rient. Dans la nuit je rentre chez moi." Duras, 97.

[20] "C'est un film français ?" "Non. International. Sur la Paix." Ibid., 65. The Japanese man, the script insists, must have "international looks." He also, curiously, speaks fluent French.

[21] "Eternelle infirmière d'une guerre éternelle." Ibid., 13.

[22] Ibid., 90.

[23] "Hi-ro-shi-ma. Hi-ro-shi-ma. C'est ton nom." "C'est mon nom. Oui. Ton nom à toi est Nevers. Ne-vers-en-France." Ibid., 124.

[24] This is said before the testimony begins, when the Japanese man insists on knowing more about Nevers and after the French woman insists on knowing why: "C'est là, il me semble l'avoir compris, que tu as dû commencer à être comme aujourd'hui tu es encore." Ibid., 81.

[25] See the *Livret* which accompanies the DVD of the film by Argos Films/Arte France Développement, 2004, page 4 (taken from an interview of Resnais published by *Cinéma 59*, No. 38).

[26] "Mon amour mort est un ennemi de la France." Ibid., 97.

[27] This is my translation of "Je désire ne plus avoir de patrie." Ibid., 114. The injustice of the treatment of women who had relationships with German soldiers during the war extended, after the war, to their children. For more information on these subjects, see Fabrice Virgili, *La France 'virile': Des femmes tondues à la Libération* (Paris: Editions Payot et Rivages, 2000), and Fabrice Virgili, "Enfants nés de couples franco-allemands pendant la guerre," 7 February 2003 (17 April 2005). <www.ihtp.cnrs.fr/recherche/enfants_franco_allemands.html>. There is also a documentary by Christopher Weber, *Enfants de Boches* (Paris: France 3, 2002).

[28] "*La Marseillaise* passe au-dessus de ma tête... C'est...assourdissant..." "La société me roule sur la tête." "Les cloches de l'église Saint-Etienne sonnaient... sonnaient..." Duras, 88, 89, 100.

[29] Born in 1914, she did not arrive in France until 1932.

[30] Jorge Semprun, *L'Ecriture ou la vie* (Paris: Gallimard, 1994). As an intellectual opponent of fascism and Nazism, he was sent to Buchenwald.

[31] After *Night and Fog* and *Hiroshima mon amour* and before *The War is Over*, Resnais tackled the subject once more with *Muriel ou le Temps d'un retour* (1963), dealing partly with torture and murder during the war of independence in Algeria.

Victims and Perpetrators:
Memory and Reconciliation in Northern Ireland

Agnès Maillot

Abstract: The conflict in Northern Ireland has resulted in the loss of over 3,600 lives. A further 40,000-50,000 people are estimated to have been injured, which translates into a large number of "secondary victims." This relatively high level of violence has strengthened the divisions between the Protestant and Catholic communities. Healing the trauma and addressing the issues of guilt and blame, and of victims and perpetrators, are essential to the conflict's resolution. In that respect, the peace process in Northern Ireland cannot be focused only on the "political" or "institutional" dimensions, but must address the crucial issue of how to deal with the past, and what mechanisms can be established in order to achieve the goal of reconciliation. The objective of this paper is two-fold: first, to analyse the manner in which the memory of the conflict has been constructed within both communities, emphasising the divergent approaches taken on fundamental issues such as the causes of the conflict, guilt and victim-hood; second, to discuss the different mechanisms instituted by governments, organisations or individuals to come to terms with the past. This leads to a discussion on the possible way forward in Northern Ireland regarding reconciliation.

Key words: Northern Ireland, War Memories, Victimization, Conflict Resolution

1.   Introduction

When Gusty Spence, former Ulster Volunteer Force member and prisoner, announced the Loyalist paramilitaries' ceasefire in October 1994, he said: "In all sincerity, we offer to the loved ones of all innocent victims over the past twenty-five years abject and true remorse. No words of ours will compensate for the intolerable suffering they have undergone during the conflict."[1] However sincerely spoken, those words were not to set the tone in the debate so fundamental to the peace process, a debate which centred on the issue of victims. Because the past is so fundamentally linked with how each community perceives not only itself but the other, the extent to which each group was prepared to review its own vision of this past was going to be tested. Nevertheless, acknowledging the role played by all participants in the conflict was considered a fundamental principle on the road to peace and reconciliation. Thirty years of what has been described as civil unrest and terrorism, or what the IRA would prefer to call a war, have deeply scarred Northern Irish society on a number of

levels. The cost of the "Troubles," as that period is more commonly
known, was high, especially given the population of the province: 3,601
people have died, of which a disproportionate number were civilians
(1,868) often referred to as the "innocent victims" of the conflict.[2] A
further 40,000-50,000 people are estimated to have been injured, which
translates into a large number of secondary victims. This high level of
violence hardened the divisions that already ran deep in Northern Ireland.
Therefore, healing the trauma, addressing issues of guilt and blame, of
victims and perpetrators, and recognizing that both sides of the conflict
had been subjected to intense suffering, constitute fundamental parts of the
peace process.

   The issue of victims is a very polarized one, obviously because of
the raw emotions that it deals with, but also because of the divide that it
generates, a divide which has both cultural and political dimensions. It is
cultural because two distinct moral visions broadly shaped by religious
affiliations are being confronted; and political because opposing positions
on the issue of victims reflect divergent views on the history and future of
Northern Ireland. There is no agreement on what happened to Northern
Ireland throughout the thirty years of violence, and no consensus as to
what term best describes the situation. "Troubles," "conflict," "unrest" and
"terrorism" are among the different labels that have been used in an effort
to avoid the term "war," the use of which was seen as tantamount to
choosing one political camp over another. In this context, the British, for
very pragmatic reasons, made strenuous efforts throughout the period to
ensure that this conflict was not categorised as a war. Citizens who
characterized the roughly 20,000 troops stationed in Northern Ireland at
the peak of the Troubles as engaged in a war could have been seen as
lending legitimacy to the "enemy."[3] This, in turn, was impossible given
that the British government and military viewed themselves in a struggle
against irregular, and inherently illegitimate, organisations.[4] Therefore,
much in the same manner as the French who refused to call the Algerian
crisis of the late 1950s a "war" for an extended period of time, the British
authorities referred to the situation in Northern Ireland as the Troubles.
The British attempt at de-politicising the violence in Northern Ireland had
some important political consequences, such as the decision in 1976 to
categorise Republican and Loyalist prisoners as criminals rather than
prisoners of war, which led to the hunger strike crises of the early 1980s.[5]

   The central role that the IRA played in Northern Ireland
throughout the conflict has undoubtedly strengthened the suspicion and
fears of the Unionist community, who tend to ascribe the responsibility for
the deaths and injuries primarily to the IRA. From the beginning of the
century, when the movement for Home Rule gained momentum, the

Unionists have regarded the demand for independence as a threat to their political and cultural survival, as their watch cry, "Home Rule = Rome Rule," clearly indicated. They fought hard to retain a link with the Crown, whereas the IRA and Sinn Féin sought full independence and the establishment of a Republic. The fact that the IRA did not accept the compromise of the 1921 Treaty - which divided the island of Ireland between the twenty-six counties known as the Free State, and the six counties of Northern Ireland - and was prepared to resort, once again, to arms only exacerbated Unionist fears that the IRA's ultimate objective represented a threat to their very survival. The various military campaigns that the IRA mounted - in 1938 in Britain, and between 1956 and 1962 in Northern Ireland - to obtain a British withdrawal from the six counties of the north of Ireland further consolidated Unionist fears, despite the fact that these campaigns were unsuccessful. The years of political violence from 1969 onward were therefore seen as a rebellion by one section of the population, namely Republicans, who were denounced for using terrorism as their main weapon and seeking to coerce the majority into a united Ireland.

The divergence in the points of view of both communities is further compounded by the fact that a sizeable fraction of the Unionist community failed to take into account the grievances of the Nationalist minority. However, Northern Ireland was undoubtedly a biased state, with an in-built Unionist majority who did not question the institutional discrimination that was instrumental to the operation of the system. Therefore, when the Civil Rights Movement emerged in the late 1960s, with simple but fundamental demands such as "one man one vote" and the end of gerrymandering, the Unionists did not seem to adequately appreciate the dissatisfaction of the Nationalist community. The tentative reforms that were proposed, together with a heavy-handed military response to the mounting political violence, sealed the fate of the Stormont parliament (the Northern Ireland parliament that was dominated by Unionists for 50 years), which was indefinitely suspended in 1972. Even today, some Unionist politicians have yet to come to terms with the dismal record of Stormont. Former Belfast Mayor and Sinn Féin MLA (Member of Local Assembly) Alex Maskey describes as a Unionist "blind spot" the fact that, according to him, most Unionists are reluctant to reappraise the part they played in creating the conditions of the conflict.

Furthermore, the Unionist community tends to make little distinction between the IRA and Sinn Féin, seeing them as being inextricably linked to each other.[6] This is fuelled by the ambiguous relationship that both sections of the Republican Movement have maintained over the years, and which is still to be clarified. Whereas Sinn

Féin leaders regularly reject charges of being members of the IRA and profess their political independence from the IRA, the fact that many Republican leaders have been accused of being members of the IRA and have spent time in prison charged with terrorist offences - even if, for some, no formal conviction was pronounced - reinforced in Unionists' eyes the belief that Sinn Féin was but the so-called political wing, the public side of the IRA. This has obvious political implications, as it explains in great part why it is inconceivable for the Democratic Unionist Party (DUP) to sit in government with Sinn Féin ministers, as long as, at the very least, the IRA has not decommissioned its weapons entirely and demonstrated with photographic evidence that it has actually done so. Ultimately, Unionists profess not to be able to trust them until the IRA has, effectively, been disbanded.

2.   Who Are the Victims?

The term "victim" is itself quite contentious. The fundamental issue relates to who should be included in this category and thus receive some form of recognition for the hurt s/he has suffered, and who should be excluded. The very fact that this question is raised shows how deeply divisive this debate is. The various definitions that have been put forward by government reports, by the Northern Ireland Commission for Human Rights and by many other organisations take a holistic view on this matter. Therefore, the category of victims includes all those who have suffered throughout the conflict, no matter in what capacity or at whose hands: "The surviving injured and those who care for them, together with those close relatives who mourn their dead."[7]

Nevertheless, many in the Unionist camp still ask whether perpetrators can be considered victims.[8] For the DUP, this is a rhetorical question, as their answer is categorically negative. But a further difficulty arises concerning who should be considered a perpetrator. Is it sufficient to blame the IRA and other paramilitary organisations for all the hurt? Or might not other forces, such as the police or the army also be considered perpetrators? Both Republicans and Unionists have clear, albeit opposite, views on this question. The former ascribe the rationale for the conflict to the occupation, as they call it, of a part of Ireland by the British, be it manifested either in political or in military terms. The latter attribute the violence to an illegitimate campaign of terrorism against a democratic state, in which there were legitimate channels through which to effect peaceful political change. In the words of Ian Paisley Jr.: "our whole community, indeed our whole country, has been the victim of the IRA for over 30 years."[9]

The statistics, however, show a different picture. According to a 1999 study on the cost of the conflict, there were, proportionally, more casualties within the Nationalist community than in the Unionist community, the death ratios being 2.5/1,000 as opposed to 1.9/1,000.[10] There are two ways in which to look at the cost of the conflict in human terms. One is to categorise the number of deaths according to the group they belonged to. This yields the following picture:[11]

Distribution of Deaths by Religion[12]

Protestants	29. 6%
Catholics	43.0
Neither	18.2
Don't know	9.2

Affiliations of the Victims[13]

Republican paramilitaries	10.0%
Loyalist paramilitaries	3.2
British Army	14.3
RUC (Royal Ulster Constabulary)	9.1
UDR (Ulster Defence Regiment	5.5
Other security forces	2.4
Civilians	53.5
Others	2.0

Organisations Responsible for the Deaths[14]

Republican paramilitaries	55.7%
Loyalist paramilitaries	27.4
British Army	8.9
UDR	0.3
RUC	1.5
Civilian	0.3
Other	0.3

Other research has focused on the circumstances in which these people were killed. Therefore, O'Leary and McGarry offer the following categorisation:[15]

Paramilitary killings of civilians	44.2%
The war between the Nationalist paramilitaries and security forces	34.8
Internecine conflict and fratricide within paramilitary organisations	6.7
The killings of Catholic civilians by security forces	5.3

The view prevalent among Unionists that the IRA is solely responsible for the 30 years of violence is, obviously, categorically rejected by Republicans. According to Danny Morrison, former Sinn Féin director of publicity,

> international opinion, the British police and the people of the twenty-six counties were assured for decades that the cause of violence in the North was the IRA campaign. If it were not for the IRA, we were told, there would be no need for the British army, for repressive powers, for state censorship. Loyalist paramilitary violence was explained away...as tit-for-tat response to the provocative Republicans, as reprisals.

In the same manner, he continues:

> Republicans, on the other hand, insisted that the campaign was an ultimate response to fifty years of Unionist misrule, then British direct rule in support of unionism. It is one of the ironies of the conflict that Unionists, the British and Irish governments and most sections of the media hid behind the IRA campaign.[16]

Therefore, for Republicans it is impossible to draw a "hierarchy of victims," which is what the Unionists tend to do in their view. Nationalists, more generally, have taken a broader view on this issue. John Hume, one of the main architects of the Belfast Agreement, and whose party has been relentless in its condemnation of violence regardless of what organisation the perpetrator belonged to, explained in an address to the Irish senate that

we cannot heal the wounds of centuries in a few years.
The violence of recent decades in particular has left deep
wounds. The hurts that have been inflicted and suffered
do not go away just because the Agreement has been
made. The Agreement cannot take away the pain, but it
is the start of the healing process.[17]

Nevertheless there is a real and tangible difficulty within the
Unionist camp regarding the issue of who is a victim. As Reverend Robin
Eames, who represents a moderate Unionist viewpoint, suggests:

Perhaps the most significant emotion to emerge within
the Protestant community as details of the *"small print"*
of the Peace Process began to emerge after the
Agreement was hurt. Certainly within the Protestant
Churches, clergy and laity gave expression to that
strange mixture of reactions which was evident in the
wider community. These included relief, uncertainty,
anxiety to push forward - but chiefly, hurt. Memories of
personal losses, death and injury surfaced and groupings
highlighted such aspects of the Patten Report on
policing as a failure to pay adequate tribute to the cost
paid by the RUC over the years of active terrorism as
well as emphasising the feelings of various victim
support organisations.[18]

3.   The Controversy Surrounding Prisoners

The difficulty of defining victims was tangible in the debate
regarding prisoners. A section of the Good Friday Agreement specifically
dealt with the early release of prisoners; at the time, this concerned around
500 men and women in Northern Ireland, the UK and the Republic of
Ireland, who belonged to Republican and Loyalist paramilitary groups.
Under the Agreement, those whose organisation was committed to, and
maintaining, a cease-fire were to be released by May 2000. This was a
fundamental point, one which greatly helped both the Republican
Movement and the Loyalist parties to sell the Agreement to their
members. But it also revealed a willingness to come to terms with past
violence and to forgive its perpetrators. The two communities reacted
differently to this measure. The Nationalists seemed to accept it as part of
the overall agreement, irrespective of the organisation to which the
prisoners belonged, whereas the Unionists found it more difficult to deal
with. A poll carried out a week before the referendum showed that the

main reason for voting against the Agreement was precisely the early
release of prisoners (45 percent of No voters), far ahead of the fear that it
could be the "beginning of a move towards a united Ireland" (18
percent).[19] Their essential belief was that a number of prisoners, no matter
what group they belonged to, were getting off lightly and that whatever
time they would have spent in prison would not sufficiently reflect the
seriousness of their crime. This measure was also seen as lending political
legitimacy to those considered to be, simply, terrorists. As academic
Duncan Murrow put it:

> While one can understand and even admire the political
> ingenuity of the solution, such compromises embody the
> hole at the centre of the agreement: everyone is left
> innocent in their own terms, while remaining guilty in
> the eyes of their opponents. We have nowhere to
> acknowledge that our relationships remain clouded by
> our experiences and perceptions of injustice and
> injury.[20]

Even though all prisoners have been released in Northern Ireland
and the high security prison of Long Kesh closed its gates in 2002, the
issue still regularly comes to the fore.[21] It is used as a bargaining tool by
Republicans and authorities alike, as in the case of the accused
perpetrators on the run (OTRs). This particular category concerns those
who have committed an offence during the Troubles but fled the
jurisdiction prior to their prosecution. This issue has angered some
Unionist politicians, who cannot accept that those who committed
offences in the past could avoid conviction. But more crucial is the issue
of whether prisoners can be considered victims. For Republicans, this was
undoubtedly the case, as the conflict generated a high number of prisoners;
Coiste, an organisation that looks after the reinsertion of former
Republican prisoners, puts that figure at 15,000 over the thirty years of the
conflict.[22] Indeed, more than half of the funding of the EU's Peace
Programme was allocated to prisoners and their representatives. A large
section of the Unionist community would undoubtedly have agreed with
the comments of *Observer* correspondent Henry McDonald:

> By the end of 2000, £4.3 million will have been
> allocated from the European Union's peace package to
> ex-prisoners' groups in Northern Ireland, while only
> £2.8 million has been spent helping their victims. The
> contrast between cash for the perpetrators and aid to the

victims will further compound the misery of those who have had to watch the killers of their loved ones and the wreckers of their businesses and property walk free under the Good Friday prisoner release scheme.[23]

The charged language used here is indicative of the raw emotions generated by this debate. For a section of the Unionist community, it is incomprehensible that those who inflicted hurt should be somehow compensated. The political argument that prisoners have something to offer to the future of the province rings hollow to those who believe the perpetrators are being rewarded for their acts. However, the fact that in 2002 the programme of government for the Northern Ireland Executive, led by David Trimble, included prisoners in the category of victims showed a willingness not to justify the acts of others but to partially take other views of the conflict into account.[24]

A number of services were put in place after the Good Friday Agreement to address the issue of victims. This followed a report commissioned in 1997 by British Secretary of State for Northern Ireland Mo Mowlam. The report, *We Will Remember Them*, was written by Kenneth Bloomfield and was to prove seminal for the future work on reconciliation. Indeed, after its publication the Northern Ireland Memorial Fund was established as an independent charitable fund "that seeks to promote peace and reconciliation by ensuring that those individuals and families that have suffered as a result of the 'Troubles' in Northern Ireland are remembered, by providing them with help and support in a practical and meaningful way."[25] There is no shortage of organisations that apply for funding. A report evaluating the Core Victims Funding Programme found that between 2000 and 2002 the Memorial Fund had financed 69 projects by 55 groups, totalling £3.1 million. Although the report does not give specific details on the organisations that applied, it does break down the successful applicants in the following way:[26]

Community	Number of projects (%)	Funds granted (%)
Cross-community	37.68	48.12
Largely Catholic	24.64	24.98
Largely Protestant	24.78	26.21
Unknown	2.90	0.69

4.   Accountability and Truth

The issues of victims and of accountability lead to another controversial area, that of the role of the police in the thirty years of conflict. In what was to generate a tense debate, a reform of the police force was envisaged by the Good Friday Agreement. Indeed, the Royal Ulster Constabulary (RUC) had been viewed by the Nationalist community as a largely partisan force, as its overwhelmingly Protestant composition clearly showed (in 1992 only 7.78 percent of its full-time members were Catholics). Moreover, the RUC was suspected of collusion with some Loyalist paramilitary groups on a number of occasions, the most obvious cases in point being the assassination of two lawyers, Pat Finucane and Rosemary Nelson. All this naturally led to Nationalist feelings of distrust and hostility towards the police. The reform of the RUC implied fundamental changes, such as its renaming as the Police Service of Northern Ireland (PSNI), the creation of new structures and the introduction of mechanisms to ensure a balance in the representation of both communities. This reform, based on the recommendations made by former Hong Kong governor Chris Patten, angered the Unionist community, which had consistently seen the role of the RUC as simply one of maintaining law and order. Therefore Unionists had little sympathy for Nationalist demands for an investigation into possible collusion between the RUC and the loyalist paramilitaries. Similarly, they saw little merit in the setting up of a tribunal of inquiry to investigate the events of Bloody Sunday, when on 31 January 1972 British army paratroopers fired on a civil rights demonstration, killing fourteen civilians.

These inquiries are of critical importance, as they touch on a fundamental part of the healing process: truth. It is the cornerstone of the peace process, for if both sides are to trust each other in political negotiations there must be honest treatment of the darker aspects of the conflict. The question of the pertinence of a truth commission, along the lines of those in Guatemala or South Africa, has been raised. Although the merits of such a measure are recognised, some drawbacks have been identified: the possibility that this could become an exercise in pointing the finger of guilt and therefore deepening antagonisms, as well as the possibility that the British side might not be willing to reveal the extent of their undercover involvement. Ultimately, however, as was pointed out by a document produced by the organisation Democratic Dialogue, "the political consensus required to establish a truth commission is lacking and likely to remain so for the foreseeable future."[27]

Getting to the truth of the actions committed by paramilitary organisations raises serious difficulties. Regardless of how well organised they may be, they are nonetheless clandestine operations hardly likely to keep files on their actions. Furthermore, a number of operations worthy of further investigation were carried out without necessarily having been approved by the commanding structures or were committed by splinter groups. Finally, to quote Democratic Dialogue again, "No truth commission can compel a paramilitary leader to attend or subpoena documentary evidence of how killings were directed; indeed, the Belfast agreement has specifically sought not to make such leaders amenable for past acts."[28] These potential difficulties were illustrated, albeit in an indirect manner, by the episode of the McCartney murder in January 2005, when a 30-year-old Catholic was beaten to death in a dark Belfast lane after a pub brawl. His assailants were members of the IRA, including some high ranking officers. Despite the campaign mounted by his sisters and partner, the complete truth of his murder has yet to be revealed. This episode showed not only the influence and intimidation that paramilitaries still exert over some communities in Northern Ireland, but also the culture of impunity in which they seem to operate and of retribution which they do not openly question, as many Republicans still refuse to cooperate with the police force.

It remains to be seen whether paramilitary organisations will cooperate with investigations into past actions. The closest instance of this taking place was in the case of the "disappeared." Throughout the conflict, but primarily in the 1970s, a number of people were killed by the IRA and buried in unidentified locations. For over twenty years, families of the "disappeared" lobbied Republicans and urged the IRA to divulge the locations of their graves. Announcing in 1999 that it had located the graves of nine people, the IRA stated:

> In initiating this investigation, our intention has been to do all within our power to rectify any injustice for which we accept full responsibility, and to alleviate the suffering of the families. We are sorry that this has taken so long to resolve and for the prolonged anguish caused to the families.[29]

In order to facilitate that process, the British and Irish governments set up the Joint Independent Commission for the Location of Victims' Remains, assuring that the information given would not be used in criminal proceedings. Gardai - the police force in the Republic of Ireland - started the search for bodies in May 1999, but as they relied on information that was not always accurate, only three of the bodies were found that year and two more in 2003.

The issue of the "disappeared" epitomizes the difficulties that the entire reconciliation process involves. Sinn Féin representatives explained that the information they relied on was, by its very nature, vague. Although this was brushed aside by their opponents as an excuse for Republicans not to come clean on this particular issue, it certainly raises the issue of the reliability of information that could be gathered from such investigations. As Martin McGuinness explains, there are no files, no records.[30] Some of those involved are dead, others have moved on, but beyond the material difficulties that this situation presents, there might be a further reason why paramilitary organisations would not be willing to embark on such a process. Finding the graves was important, but the next step was, necessarily, understanding why these people were disappeared and explaining their execution. Margaret McKinney, whose son Brian disappeared in 1978, expressed her distress when she asked: "What use are apologies to me? I want peace. I want to know why he was murdered." [31]

Although it took the IRA a number of years to publicly apologise to all relatives of non-combatants killed in their operations, which it did in July 2002, Gerry Adams addressed the Unionist community early in the process in the following terms:

> I'm not going to ask you to forget the past nor to forgive
> Republicans for the pain we have visited on you. At the
> same time I don't expect Nationalists or Republicans to
> forget what you inflicted on us. However, the wrongs of
> the past must not paralyse us. We must not be trapped in
> a web of suspicion and doubt about each other.[32]

It remains to be seen how far down the road of introspection all sides are prepared to go. The reticence shown by the British government to make public the contents of the Cory report on the murder of Pat Finucane is indicative of the difficulties involved. The fact that a sizeable proportion of Unionists are dissatisfied with the manner in which the issue of victims is being dealt with is another obstacle to reconciliation. Finally, the continued criminal activities of all paramilitary organisations, including the IRA, are hindering the process of transparency and accountability. The way forward might lie in the acceptance that - whoever the perpetrators may be, whatever vision of the past individuals and communities might have - all sections of the population have been affected by thirty years of violence, and none can claim a monopoly of pain. In the words of Kenneth Bloomfield: "We truly need to remember those who have suffered, to grieve at the side of this communal grave, to reflect upon the truth of what occurred and to move forward from there."[33]

Notes

[1] Peter Taylor, *Provos: The IRA and Sinn Féin* (London: Bloomsbury Press, 1997), 348. The Ulster Volunteer Force is one of the two main Loyalist paramilitary groups, the other being the Ulster Defence Association

[2] Marie-Therese Fay, Mike Morrissey and Marie Smyth, *Northern Ireland's Troubles: The Human Costs* (London: Pluto Press, 1999), 164.

[3] Despite the peace process some 13,000 troops still remain. Northern Ireland Office, *Responding to a Changing Security Situation* (Belfast: HMSO, 2003).

[4] Aside from the Provisional IRA, there were a number of other Republican paramilitary organisations throughout the Troubles, such as the Official IRA, which declared a ceasefire in 1972; the Irish National Liberation Movement, which was created in 1976 and was active throughout the conflict; the Real IRA and the Continuity IRA, both of which were funded as a result of dissent among the ranks of the IRA over the Peace Process and the general direction that the Republican Movement was seen to be taking (such as the signing of the Good Friday Agreement and the acceptance of decommissioning as part of the process).

[5] The term "Republican" is used here to refer to the members of the Republican movement, that is, Sinn Féin and the IRA, as well as their supporters, and should therefore be distinguished from the broader term "Nationalists," which encompasses all members of that community whose political allegiance is divided, broadly speaking, between the Social Democratic and Labour Party (SDLP) and Sinn Féin. The term Loyalist is used for the members of paramilitary organisations of Unionist obedience, such as the UDA (Ulster Defence Association) or UVF (Ulster Volunteer Force).

[6] In Unionist rhetoric the term "Sinn Féin-IRA" is commonly used to express the perceived amalgamation of the two groups.

[7] Kenneth Bloomfield, *We Will Remember Them* (Belfast: HMSO, 1997), 14.

[8] The Unionist camp is divided between two main parties, the Ulster Unionist Party led by David Trimble and the Democratic Unionist Party of Ian Paisley. The former has been more accommodating towards Republicans, and towards Sinn Féin in particular, accepting that any political solution had to involve all shades of opinion and agreeing, albeit after much internal turmoil, to form a coalition government for Northern Ireland which included two Sinn Féin ministers. The DUP, on the other hand, has maintained a tougher line regarding Republicans, deeming them

unacceptable in any democratic process as long as the IRA remains in existence. It has refused to sit on governing bodies which included representatives from Sinn Féin. This line has been to some extent espoused by a great number of Unionist voters, as they increasingly saw the failure of total decommissioning of weapons and the continued activities of the IRA as major obstacles to the political process. Therefore, at the last British general election in May 2005, the DUP obtained a resounding victory over its Unionist opponent, taking 9 of the 18 seats that represent the province at Westminster, leaving the shattered UUP with only one seat (the share of the vote went from 22.5% to 33.7% for the DUP, against 26.8% to 17.1% for the UUP). On the other side of the spectrum, Sinn Féin has successfully transformed its political progress into votes, managing to outperform its major rival, the SDLP, at the 2005 elections, with a share of 24.5% of the vote, up from 21.7% for the previous general election, while the SDLP saw its vote decline from 21.0% in 2001 to 17.5% in 2005). This leaves the smaller parties, such as the Alliance Party of Northern Ireland, with a vote averaging 3.5 to 4% and no representation in parliament.

[9] Northern Ireland Forum Debates, 4 July 1997 (19 July 2005). http://www.ni-forum.gov.uk.debates.htm.

[10] Fay, 168.

[11] Ibid.

[12] Ibid., 169.

[13] Ibid.

[14] Ibid.

[15] Brendan O'Leary and John McGarry, *The Politics of Antagonism* (London: Athlone Press, 1997), 28.

[16] Danny Morrisson, *The Cause of Violence*, n.d. (19 July 2005). <http://www.dannymorrison.com/articles/causeofviolence.html>.

[17] John Hume, Address to Senead Eireann, 3 March 2004 (19 July 2005). <http://www.sdlp.ie/prhumeseanadeireann.shtm>.

[18] Robin Eames, "The Religious Factor" in *Protestant Perceptions of the Peace Process in Northern Ireland*, ed. Dominic Murray 2000 (19 July 2005). <http://cain.ulst.ac.uk/events/peace/murray/eames.htm>.

[19] Deaglan de Breadun, "Unionist Support for Agreement drops sharply, poll shows," *The Irish Times*, 15 May 1998 (19 July 2005). <http://www.ireland.com/newspaper/front/1998/0515/archive.98051500002.html>.

[20] Duncan Morrow, "Forgiveness and Reconciliation," *Democratic Dialogue,* Report 13, 2001 (19 July 2005). <http://cain.ulst.ac.uk/dd/report13/report13a.htm>.

[21] There are still four IRA prisoners in the Republic of Ireland who were convicted for the killing of Garda Gerry McCabe in 1996. The Irish government considers that their case is not covered by the Good Friday Agreement, and this has given rise to several public controversies with Sinn Féin, which maintains that all prisoners should be released.

[22] See the organisation's website at http://www.coiste.ie/.

[23] Henry McDonald, "Listen to the Victims," *The Observer*, 29 October 2000 (19 July 2005). <http://observer.guardian.co.uk/comment/story/0,,389520,00.html>.

[24] "Consultation Paper on a Victims' Strategy" (19 July 2005). <http://www.victimsni.gov.uk/victimsstrategy/chapter1.htm>.

[25] *Northern Ireland Memorial Fund Website* (19 July 2005). <http://www.nimf.org.uk/about_us.htm>.

[26] Clio Evaluation Consortium, *Evaluation of the Core Funding Scheme for Victims'/Survivors' Groups* (19 July 2005). <http://www.brandonhamber.com/projects_clio.htm>.

[27] "Executive Summary," *Democratic Dialogue*, Report 13, 2001 (19 July 2005). <http://cain.ulst.ac.uk/dd/report13/report13a.htm>. "Democratic Dialogue was initiated as a think tank to address the continuing challenges of political accommodation which the ceasefires in themselves did not resolve, as well as the social and economic issues long neglected by Northern Ireland's single-item political agenda." Democratic Dialogue (viewed 19 July 2005). <http://www.democraticdialogue.org/history.htm>.

[28] "Executive Summary," 19 July 2005.

[29] Clare Murphy and Maol Muire Tynan, "IRA says it has located burial places of nine of its victims," *The Irish Times,* 30 March 1999 (28 May 2005). <http://www.ireland.com/newspaper/ireland/1999/0330/archive.99033000045.html>.

[30] Martin McGuinness, interview with author, Derry, December 2003.

[31] *The Irish Times*, 5 October 2004.

[32] Gerry Adams, *Presidential Address to Ard Fheis*, 1996 (19 July 2005) <http://sinnfein.ie/peace/speech/3>.

[33] Bloomfield, 23.

Part III

Cyberwar

E-Jihad, Cyberterrorism and Freedom of Speech

Markku Jokisipilä

Abstract: The information revolution has given the concepts of netwar and cyberwar growing importance. Various activist and extremist groups have successfully used the Internet and cellular phones to recruit, to coordinate and to disseminate propaganda. Together with network-style organizations, new technologies allow non-state actors to pose unprecedented challenges to state power. Although these new challenges are mostly nonmilitary, they have profound implications not only for warfare but also for national sovereignty and international law. The costs of escalation in cyberwar are small compared to traditional warfare, but countering the threats presented by network opponents requires transnational and cross-jurisdictional cooperation, not easily accomplished by state hierarchies. Local cyberwars carry an inherent risk of spreading throughout the Internet in the form of viruses and worms. They can also damage the real world by shutting down electrical grids, telephone service, air traffic control systems and other essential civil and governmental information networks.

Keywords: Cyberterrorism, Cyberwar, Internet, Palestine, Israel, Free Speech

1.   Introduction

On 7 October 2000 a team of Hezbollah fighters illegally crossed the Israeli-Lebanese border and kidnapped three Israel Defence Force (IDF) soldiers. As Israel, in retaliation, threatened to bomb Beirut and Damascus, the IDF and Jewish settlers unleashed their rage against the Palestinians. Reaction also spilled over to cyberspace, as a group of teenage Israeli hackers attacked the websites of Hezbollah, the Palestinian Authority and Hamas. The Palestinian hacker community responded with calls for "electronic Jihad" and the massive targeting of official Israeli websites. During the course of this cyberconflict, or *Interfada* as it was dubbed, hundreds of websites on both sides were continuously attacked. All online activities were seriously hampered and the whole Middle Eastern electronic infrastructure came under heavy strain. As foreign volunteers joined in and new targets were constantly hit, conflict spread across the globe. Although the Interfada has gradually abated, e-Jihad rages on, making cyberspace one of the most important arenas of ideological arm wrestling between the Islamic world and the West.[1]

Many observers saw the Interfada as an online variant of "the clash of civilisations" in a true Huntingtonian sense: the first full-scale cyberspace war. The success of the pro-Palestinian side was seen as a glimpse of things to come, the growing validity of e-Jihad and the emergence of political insurgency relying more and more on new information technology (IT). Moreover, the overall course of events conformed to the general patterns of cyber warfare presented in theoretical literature. In a cyberwar it is essential to obtain a global support network. Because the costs of escalation are low, cyberconflicts escalate and internationalise rapidly. Within the first weeks of the Israeli-Palestinian conflict, Pakistani, United States, Brazilian and Chinese hackers had jumped in, and by the end of 2000 ideologically motivated international participation reached levels comparable to the Spanish Civil War (1936-1939). The battlefield expanded as American, Iranian, Lebanese and Syrian websites came under attack.[2]

The Interfada highlighted the current shift in organisational paradigms from hierarchies to networks. Hacking and other cyberattacks are textbook examples of new asymmetric threats that a state computer system would find difficult to prevent and to justifiably retaliate against. Traditional hierarchies like the Israeli government and IDF could not counter the offensive of decentralized flexible networks, which have neither hierarchical structure nor a single leader and which rely heavily on IT. Traditional means of state power are simply inadequate for blocking cyberattacks by international hacker networks. Successful cyber warfare necessitates transnational and cross-jurisdictional cooperation and active use of networked styles of organisation, all very difficult for states to match or interdict.[3]

Although the highly publicized scenarios of cyberterrorism inflicting damage on the physical world - by tampering, for example, with electricity and telecommunications networks or air traffic control systems - have not been realized, cyberwars can be very pernicious. They can spread mischief throughout the Internet in the form of viruses and worms, effectively jamming all online operations for long periods of time. The Interfada displayed the potential of organized computer hacking to cause serious economic and governmental damage better than any other cyberconflict before or since. At the same time, however, it also illustrated the dangers of blurring the boundaries between legitimate online protest and unjustifiable acts of electronic vandalism or outright cyberterrorism.

Protecting critical national infrastructures in the information age contains a huge risk of governmental overreach and is potentially very detrimental to online freedom of speech and cyber civil rights. Exaggerating the threat posed by terrorist use of IT generates over-the-top technological anxieties and obscures the huge positive potential of the Internet. This is all the more obvious after the 9/11 terrorist attacks on the United States and the subsequent global war on terror. A declaration by Reporters Without Borders articulates this:

> On 10 September 2001, the Internet was still a place of hopes and dreams that was going to give everyone access to impartial information and undermine dictatorships. A few days later, it had became [sic] a lawless place where Al-Qaeda had managed to plan and coordinate its attacks. The Internet began to frighten people. The 10 September was the last day of a golden age of free online expression. Since then, Big Brother has loomed ever closer.[4]

2.   Revolution of Information and Networks

Four basic forms of social grouping underlie the organisation and historical evolution of all societies: kinship-based tribes, hierarchical institutions, competitive-exchange markets and collaborative networks. Although incipient versions of all four forms have been present since ancient times, they have gained strength at different rates and matured in different historical epochs. Networks seem to be the dominant form of social organisation for the future. In the coming years we will be witnessing more frequent low-intensity conflicts waged by networks against hierarchies and institutions, such as, for example terrorist organisations or ethnic factions versus states. The nature of these conflicts covers all areas of human life, from military operations to social movements to economic market relations.[5]

The revolution of computerized information and communication technologies, accompanied by innovations in management and organisation theory, has emphasized the strategic centrality of information. It has dramatically boosted efficiency, shaken up social systems, and transformed conventional models of thinking and operating. It has eroded the position of traditional hierarchies by diffusing and redistributing power, often empowering the previously weaker and smaller parties. As real-time interaction, even over great distances, has become a reality, the significance of geographical location and national boundaries has diminished, producing radical changes in people's spatial and temporal horizons. As many jurisdictions and responsibilities have been redefined, closed systems have been forced to open. The IT revolution is favouring the growth of networks by enabling diverse and dispersed actors to co-operate, and by supplying them with better information than ever before.

Many institutions have supplemented or even replaced hierarchical organisations with flexible, network-like structures.[6]

The importance of information has also multiplied in military strategies and warfare. Information warfare has ceased to be merely an auxiliary to the search-and-destroy tactics utilized by conventional forces. The ability to disrupt an enemy's information systems is central in all present-day military conflicts. As the coiners of the terms netwar and cyberwar - John Arquilla and David Ronfeldt of the RAND Corporation - have pointed out, warfare is no longer primarily about who has the most capital, manpower and technology, but about who has the best information concerning the battlefield. They use an analogy of a chess game where you see the entire board, but your opponent sees only his own pieces. However, the growing reliance on IT by advanced societies is a double-edged sword, forming a potential strategic vulnerability to a nation's critical infrastructure. This makes the use of new technologies tempting for weak states and non-state entities, which would not stand a chance in traditional symmetric force-on-force combat against stronger state enemies.[7]

The IT revolution has given birth to a completely new kind of battlespace dynamics. Omnidirectional weaponry, such as computer viruses or radio frequency weapons, create possibilities for mass disruptive targeting, which could bring technology-dependent Western societies to the brink of chaos without causing the high levels of physical destruction associated with traditional weapons of mass destruction. On the other hand, the emergent post-modern military paradigm, with its temporal and spatial breakdown of the modern battlespace, greatly extends the field of possible military targets. Arquilla and Ronfeldt saw the 9/11 attacks as definitive proof that al-Qaeda and other international terrorist organisations are developing a new war paradigm based on swarming tactics, that is, striking multiple targets simultaneously from multiple directions.

At the organisational level this means major confrontations between state-level hierarchies and non-state networks. Ultimately, changes wrought by the IT revolution could shake the very foundations of the world-system of nation-states. National military forces will have to develop effective defences against entirely new kinds of attacks and eventually be able to engage in similar offensives themselves. Most importantly they will have to develop a new field of information strategy.[8] The superiority of networks lies precisely in their unequalled ability to gather, use and process information, made possible by their flat hierarchies, decentralized authority and loose lateral ties. The development of new kinds of military units, networked structures and disruptive targeting capabilities might well prove indispensable in order for states to successfully combat diverse informal networks.[9]

The Mexican Zapatista movement (EZLN, *Ejército Zapatista de Liberación Nacional*) has been seen as the first representative of a new, non-hierarchical model of political radicalism. It was started in 1983 as a classic Marxist national liberation movement, but transformed during its first decade of existence into a completely novel type of guerrilla organisation, relying on advanced communications technology and a global network of supporters. Since the 1980s, traditional hierarchical groups in the Middle East that were pursuing nationalist or Marxist agendas, like the Abu Nidal Organisation and the Popular Front for the Liberation of Palestine, have been challenged by Islam-motivated groups with more elastic organisations. For example, Hamas and Al-Qaeda are flexible networks, consisting of loosely interconnected semi-independent cells and lacking a single hierarchical chain of command. The IT revolution has greatly enhanced the operational environment of these groups by supplying them with cheap, quick, secure and multifaceted flows of information and finances, and they rely heavily on the new technology. For example, Hezbollah has its own special Internet department administrating a multitude of websites in Arabic, English and Hebrew. Similar trends of networked organisation and IT use can be observed in most of the new social and political activist movements.[10]

3.   Netwar, Cyberwar and Hacktivism

Although the change brought about by technological and organisational revolutions is in an embryonic phase, and information-age ideologies are only just starting to take form, some general observations can be made. The rise of networks will lead to the adoption of netwar strategies by diverse groups, whether social movements, terrorist organisations or ethno-nationalist factions. The term netwar, often used interchangeably with infowar, cyberwar or Third Wave warfare - though not precisely the same thing - was conceived by Arquilla and Ronfeldt in the early 1990s. It refers to "societal-level ideational conflicts waged in part through Internetted modes of communications," where one or both sides are trying to disrupt, damage or modify a target population's knowledge about itself and the surrounding world by various diplomatic, propagandistic and psychological measures, political and cultural subversion, media deception or interference, computer network infiltration and online promotion of dissident or opposition movements. Netwars take place mostly on the non-military end of the conflict spectrum, but they can overlap with or turn into more militarized cyberwars. They may occur between nation-states, between governments and non-state actors, as well as between rival non-state actors. During the past ten to fifteen years the rivalry between non-state networks has continuously gained in importance and is likely to do so in the future. Nation-states will likely not disappear, but they will certainly be transformed by these developments.[11]

Cyberwar is a subset of information warfare that involves actions taken in the virtual reality of computer networks, especially the Internet. It refers to information-related operations aiming to disrupt and/or destroy an opponent's critical information and communications systems, and possibly even parts of its physical infrastructure. Successful cyberwar requires that one party gather all possible information about its adversary while maintaining its own anonymity, and using this knowledge to expend minimal capital and manpower in the struggle. Cyberwar usually involves intensive use of IT to interfere with an opponent's information systems, but it is not only about technology. It also means organisational adaptations and new institutional designs for more networked and decentralized structures and calls for a new understanding of force requirements and tactical methods. In some cases these networks might only be countered by other networks when the necessary flexibility, lateral connectivity and cooperation across institutional boundaries cannot be achieved any other way. In the views of Arquilla and Ronfeldt, cyberwar transforms the nature of warfare so radically that it can be seen as a twenty-first century equivalent of blitzkrieg.[12]

The Internet can be used as an auxiliary tool in asymmetric warfare against a nation-state. The more networked a group or movement is structurally, the more inclined it is to make extensive use of IT. New technological innovations give the weaker side - whether a social movement, a terrorist group, a rogue state or an individual hacker - a chance to compete against state opponents, which in terms of conventional military and economic power enjoy overwhelming superiority. Theoretical literature distinguishes three principal ways of using IT to further a social, ideological or political cause: activism, hacktivism, and cyberterrorism. Activism refers to normal, non-disruptive use of the Internet for political influence: browsing the Web for information, building websites, posting messages, sending e-mails or participating in online discussions in order to debate issues, to form coalitions and to plan and coordinate activities. Electronic civil disobedience or mass virtual action are not deliberately disruptive, although they often seriously interfere with or completely hinder the normal operation of a target website. Usually this happens through virtual sit-ins, whereby thousands or even tens of thousands of people converge on a certain website at a given time, or by otherwise blocking its normal functioning without intruding on the site or altering its contents in any way. Cyberconflicts provide many good examples of swarming as an emergent form of tactics, that is, electronic converging of dispersed groups spread over great distances on a single target.

Hacktivism refers to the crossing of computer hacking with activism, wherein the hacktivist uses the Internet to disrupt a target's website or otherwise interfere with its operations. Hacktivists may disturb the target up to the point of complete standstill yet they do not aim to cause any serious damage. Cyberterrorism refers to politically and ideologically motivated efforts to use the Internet as a vehicle for causing serious "real world" damage, such as the destruction of critical physical infrastructure or the loss of human life. Penetrating a nation's air traffic control system and causing planes to crash or to collide in mid-air is an often cited example of cyberterrorism. It is unclear whether online aggression resulting only in economic damage, however severe, should be regarded as terrorism. On the whole, the boundaries between activism, hacktivism and cyberterrorism are not clear-cut. An act considered by some as activism may well be qualified by others as cyberterrorism.[13]

Typical cyber warfare tactics include website defacement; denial of service attacks; domain name service attacks; use of worms, viruses and Trojan horses; exploitation of inherent computer security loopholes and unauthorized intrusions into an opponent's computer systems and networks. Defacement means the disruption of a website so that it no longer performs the function for which it was designed. It usually includes altering the contents of the page by placing on it, for example, propaganda, profanity or pornographic images. Domain name service attacks aim to secretly redirect traffic from one website to another, usually to one on the attacker's own server. It enables the attacker to disseminate false information and mislead web users. A denial of service attack floods the victim's system with requests for information to such an extent that the system is no longer able to operate. In February 2000, this kind of attack closed down many high-profile commercial sites for days, including Amazon.com, CNN.com and Yahoo. Worms, viruses and Trojan horses are malicious autonomous computer programmes which aim to disrupt and even destroy the targeted systems or to build in secret back doors for later access. Unauthorized intrusion is a hostile takeover of a computer system by an outside actor. The degree of intrusion can vary from a relatively modest "peeking" at individual files to root access penetrations through which the attacker gains complete control of the system.

4.   Seminal Effect of the Internet

As a technical innovation and tool of communication the Internet is potentially more powerful than anything we have seen before. In principle it can be accessed and used by anyone at any time from anywhere. In 2004 the worldwide Internet population amounted to 1.08 billion people, and the projected figure for 2010 is 1.8 billion.[14] The World Wide Web transcends national boundaries and enables individuals and groups of people from anywhere in the world to communicate globally. Technologies like streaming media and web-casting combined with advanced multimedia applications have further enhanced the Internet's capability for spatial contraction. The Web provides many venues for discussion and debate, such as e-mail, newsgroups, forums and chat rooms. It has changed the landscape of political mobilization by energizing and empowering grassroots political activities of every stripe, and by providing activists a new forum for protest and a new means of conducting cyber-insurgencies. Opposition groups and dissidents are forging alliances and coalitions using mailing lists, e-mail alerts, petition sites, sample letters to government and civil authorities, virtual sit-ins and e-mail bombing. There are many spectacular examples of even more offensive use of new IT in recent years. The Kosovo conflict in 1996-99 is said to have been the first war of the Internet. It turned cyberspace into "an ethereal war zone where the battle for the hearts and minds is being waged through the use of electronic images, online discussion group postings, and hacking attacks."[15] In January 2001, Philippine activists used cell phones and the Internet to coordinate huge demonstrations against then-President Joseph Estrada. In April and May of the same year, hundreds of Chinese hackers launched a cyberattack against the United States after an American reconnaissance aircraft collided with a Chinese interceptor. In the United States presidential elections of 2000 and 2004, supporters of Ralph Nader conducted a massive online vote-swap with supporters of Al Gore and John Kerry in other states.[16]

The nature of the Internet as a borderless, fast, atomised and anonymous network allows it to elude the grip of traditional means of state power. Advanced encrypting tools enable secure transmission of information and make the monitoring and controlling of electronic traffic very difficult. The Internet has clearly tilted the political balance of power in favour of the individual and, in particular, has made repressive governments less capable of controlling and censoring their citizens. Once any information is posted on the Web, it will in all likelihood be accessible somewhere as an archived copy or mirror site even if the original site is shut down. As a result, state structures are more exposed to opposition, protests and dissident actions. Unlike traditional mass media, the Internet can be exploited by individuals and small groups with few resources virtually to the same degree as by large and richly-funded coalitions and organisations, thereby fostering the democratic process. The most enthusiastic free speech proponents have even saluted the Internet as a harbinger of online global democracy.

As the massive reliance on IT within advanced countries has facilitated daily life in countless different ways and further increased those nations' economic superiority, there is a flipside to the coin as well. The combination of wiredness and openness in high-tech societies creates many new possibilities for both cyber-banditry and politically motivated cyberattacks. It has been estimated that in 2002 alone the worldwide business losses resulting from cyberattacks were around US $1.5 trillion.[17] Especially in the aftermath of 9/11 people have become aware of the various threats - some of them extremely sinister - that technology-dependency has generated. The same technology that is used to pursue good can also be easily exploited by those with harmful intentions.

Already a decade ago some researchers envisioned that cyberspace would be the primary battleground of the future. Access to information today is just as crucial as possession of petroleum or military hardware. The boundaries between military and non-military conflicts are becoming all the more unclear and they are also deliberately blurred. In particular, military strategists see that IT could provide adversaries of the United States a way around its military superiority by enabling them to launch attacks on critical American infrastructure from anywhere in the world and bring any distant conflict directly to America's heartland.[18] Although serious efforts to wage offensive information war have so far been virtually non-existent, there is a growing fear of large-scale cyberterrorism, such as the use of electronic networks by diverse militant groups to inflict physical destruction to further their ideological cause.[19]

The possibility of serious cyberattacks against vital national infrastructures has been a subject of debate for many years and especially after 9/11. Dreadful scenarios of cyberwars expanding to the physical world have been suggested: terrorist attacks paralysing electrical power grids, air traffic control and rail networks, banking and finance, gas and oil supplies, emergency services, governmental functions and so forth. Richard Clarke, a former White House terrorism and cybersecurity advisor, has been continuously warning of a "digital Pearl Harbour": computer-based attacks causing massive destruction and loss of life.[20] According to James Adams, internationally renowned expert on computerized warfare and CEO of the web security firm iDefense, we are witnessing

> a relative replication of the escalatory ladder, translated
> into the virtual space. Malicious hackers have become
> the guerrilla fighters, and massed armies have taken the

form of cybersoldiers representing organisations that are
criminal, terrorist and governmental in nature.[21]

Because the move towards more networked structures in the
organisation of terrorist groups resembled and coincided with
developments in cyberspace, it has been compelling to draw analogies
between terrorism and the Internet. Some analysts have even presented
them as symptoms of the same phenomenon, the perilous potential and
inherent danger of new technologies. Western news media have for years
spoken of Middle Eastern techno-terrorism exploiting the Internet, e-mail,
cell phones and satellite communications to coordinate their activities and
promote their ideology. The "Internetted organisation" of the terrorist
groups has been underlined; al-Qaeda, for example, has been referred to as
"a kind of terrorists' Internet."[22] The militant wing of online activists has
further contributed to this kind of thinking; for example, a representative
of Electronic Disturbance Theater, a group of American electronic civil
disobedience pioneers, stated the following:

> We do not believe that only nation-states have the
> legitimate authority to engage in war and aggression.
> We see cyberspace as a means to enter present and
> future arenas of conflict, and to do so across
> international borders.[23]

The fear of cyberaggression is not completely without
justification. Compared to traditional physical methods employed by
terrorists, cyberterrorism does have some unique advantages. It can be
conducted remotely, anonymously, cheaply and without immediate risk to
the perpetrator's life. Moreover, a successful act of cyberterrorism would
be likely to receive extensive media attention.

After 9/11 there was a clear trend in the United States and other
countries to use the national emergency and the threat of terrorism as a
justification to suppress various forms of protest and as a pretext for
extending electronic surveillance regimes: many anti-globalisation protest
groups, which had no connection whatsoever to the strikes in New York
and Washington, were reclassified as terrorist organisations. A post-9/11
advisory issued by the FBI's National Infrastructure Protection Centre
stated the following:

> While the cyber damage thus far has been minimal, the
> infrastructure will certainly be a target of cyber
> protestors and hacktivists in the future, with the

potential goal being intentional destruction rather than
public embarrassment or purely political statements.[24]

Accounts like this exaggerate the destructive potential of these new tools.
Malicious hacking can damage electronic infrastructure and even cause
physical harm, but seizing control of computer systems from the outside
requires exceptional skills and inside knowledge. It is still far easier to
bomb a target than to hack a computer inside it.

It is extremely worrisome that many governments have levelled
serious accusations against the proponents of Internet freedom of speech.
British Foreign Secretary Jack Straw has underlined that those who resist
stronger Internet controls are actively hurting the fight against terrorism.
There seems to be a general trend among state authorities to deliberately
lump together all kinds of online dissidence, whether criminal or
emancipatory.

Because cybercriminals rely exactly on the same technology as
law-abiding citizens, it is vital to distinguish between legitimate online
debate and protest on the one hand and criminal activities on the other.
According to a useful definition by Mark M. Pollitt, the chief of the FBI
Computer Analysis Response Team, cyberterrorism is "the premeditated,
politically motivated attack against information, computer systems,
computer programs and data which results in violence against non-
combatant targets by subnational groups or clandestine agents."[25] If a
cyberattack does not affect the physical world in any significant way, it
cannot be considered as terrorism, and there is little ground for
governmental interventions or other heavy-handed countermeasures. Even
large-scale cybervandalism, although undeniably a nuisance, does not
qualify as cyberterrorism. This should be taken into account in assessing
the real threat posed by diverse cyberactions. Excessive and ill-conceived
governmental online policing would endanger the legitimate claims for
privacy and freedom by ordinary citizens.

Like other groups, terrorist organisations use IT for internal
purposes such as command, control and communications, but so far at
least they have been unwilling or unable to engage in any ambitious
offensive information warfare operations.[26] Disruption is one thing,
destruction completely another. Achieving electronic attack capability and
the ability to circumvent protective mechanisms requires a high level of IT
skills and equipment. In the cyberworld, producing the desired effect with
a certain device or software is much more difficult because of the
technical complexities involved. Because terrorists want to be sure that
their weapons work, they will most likely stick to their traditional arsenal.

Moreover, cyberattacks rarely if ever produce the kinds of dramatic, fear-arousing, media-attracting effects that traditional terrorist strikes do.

As more effective defensive measures are developed, terrorists will have to upgrade their skills and tools constantly. This of course will mean expending huge amounts of time, money and organisational resources. Considering the breathtaking speed of IT development and the fact that bombing still produces the expected results, it is hardly surprising that most terrorist groups prefer their traditional tactics. Although disruptive cyberattacks are easier to conduct, they lack immediate and visible physical-world damage. Terrorists want to make their enemies fear for their lives, and this is next to impossible to achieve with electronic sabotage alone.[27]

The much-repeated forecast from 1991 that "tomorrow's terrorist may be able to do more with a keyboard than with a bomb" has not been realized, and this seems unlikely to change in the foreseeable future.[28] Cyberattacks against critical infrastructure by "ordinary" hackers motivated by thrill-seeking, fame or financial gain are even less likely. Because almost any hack tends to arouse at least some public interest and notoriety on the Web, most hackers choose their targets based not on their strategic or infrastructural importance but on their vulnerability.

5.   Israeli-Palestinian Cyberconflict

The first actions in the Israeli-Palestinian cyberconflict had been taken in 1999, but it was only after the outbreak of the Second Intifada in September 2000 that it moved to a high-intensity level. The parties had been engaged in a fierce online propaganda war since the mid-1990s, with both official Israeli institutions and major Islamic websites bombarding online audiences with their versions of the Palestinian question. This struggle has been far more evenly matched than the real world conflict, largely because of the vast international online support from Palestinians. In particular, a website called *Electronic Intifada* (EI), launched by two Palestinians and two foreign pro-Palestinian human rights activists, has been able to supplement mainstream Western representations of the conflict by offering the international online community documented and fact-checked information from a Palestinian point of view. Today EI is one the world's most followed resources on the Israeli-Palestinian conflict, visited by millions each year and used as a frequent source of information by international journalists.

After the kidnapping of three Israeli border patrol soldiers by Lebanese Shiite fighters was reported in the news on 7 October 2000, a group of Israeli teenagers penetrated a Hezbollah website and placed on it an Israeli flag accompanied by a recording of the Israeli national anthem.

Soon after, there were numerous separate chain e-mails circulating in Israel offering instructions on how to strike against pro-Palestinian and Arab websites. Israeli hackers founded a special website inciting volunteers to join in and providing them with clickable menus of targets to attack. Within a few days the Hezbollah website crashed under millions of service requests. One hacker broke into a Palestinian Authority database and published cell-phone and fax numbers of Palestinian leaders on his own website. When the Palestinian side got back on its feet after the initial shock it began a sustained and coordinated retaliation by defacing and eventually crashing several top-level Israeli domain websites, including those of the Israeli Parliament (Knesset), the prime minister, the IDF, the foreign ministry, the Bank of Israel and the Tel Aviv stock exchange.[29]

A few more detailed examples will suffice to outline the nature of the cyberclash. The original graphics on the website of Jerusalembooks.com, the largest online bookstore in the Middle East, were replaced with the word "Palestine" and the text asked if the Torah taught Israelis to kill children and rape women. The e-mail inbox of the webmaster of the Israeli Foreign Ministry site was flooded with a tool called QuickFire, which with one click of a mouse sends 32,000 e-mails to the victim. A Pakistani hacker with the nickname Doctor Nuker replaced the website of the American Israeli Public Affairs Committee with another criticizing the harsh measures used by the Israeli government and denigrating Jewish people in general. He then broke into the committee's databases and seized personal information on 700 members of the committee, including telephone numbers and credit card information, and posted it on a website for the entire world to see.[30]

As the stakes grew, UNITY, an online Muslim extremist group with links to Hezbollah, laid out a four-phase strategy for destroying Israel's Internet infrastructure. In phase one, official Israeli sites were attacked, phase two incorporated hacks against major financial sites, phase three involved targeting the Israeli Internet and telecommunications infrastructure, and finally phase four aimed at bringing down the biggest e-commerce sites, costing millions in transactions. Attacks in all four phases were evident during the first two months of conflict, and it is believed that successful hacking of large commercial sites caused an eight percent dip in the Tel Aviv Stock exchange.[31]

Like all previous major cyberconflicts, the Israeli-Palestinian electronic skirmish soon internationalised to involve politically motivated activist hackers from dozens of nationalities on both sides. As Palestinians were receiving help from the global online Muslim community, Israeli hackers, running out of Palestinian targets and backed up by allies from the United States, turned their swords against pro-Palestinian and Islamic websites in Iran, Lebanon, Syria and the United States. Brazilian hackers, who wanted to show off their skills, subsequently attacked both Israeli and Palestinian sites. The spectrum of participants encompassed everyone from average Internet users to traditional terrorist groups to purely a-political hacker teams. Ben Venzke, director of iDefense, summed it up well: "We're starting to see groups that have no connection or relationship to anything going on in the region jumping into the fray because they think it's a neat thing and want to be a part of it."[32]

Tactical and technological innovations were not long in coming. The ringleaders of the conflict recruited thousands of less technically qualified, but nonetheless ideologically motivated foot soldiers to provide the critical mass. "Exploit scripts," computer programmes used in attacks, were widely and rapidly disseminated, and hostile codes were frequently rewritten by the opposing side and then flung right back as modified versions. It has been estimated that the Interfada was run by only thirty hackers, who then fed a large and international flock of "script kiddies" with the necessary tools to scan the other side's vulnerabilities and attack them. A group of self-described ethical hackers known as Israeli Internet Underground (IIU) set up a website to help Israeli sites defend themselves against the attacks and joined ranks with an Internet security company called 2XS Ltd, which promised to provide Israeli website administrators with protection tools and security advice.

High-intensity conflict raged for months and accurately mirrored events in the real world. There were bouts of frantic cyber activity immediately following suicide bombings and IDF artillery barrages. Over the course of the next two years, hundreds of websites on both sides were subjected to distributed denial of service attacks, website defacements, misinformation campaigns, system penetrations, Trojan horses and viruses. What had started out as a technical cock-fight between highly skilled individual hackers evolved into a full-scale political cyberwar involving thousands of people on both sides. Spectacular shows of hacking expertise were followed by floods of electronic expressions of racism and other obscenities.[33]

The electronic infrastructure in the region was under great strain as the attacks and counterattacks shut down popular information and financial sites for days or otherwise disrupted the normal functioning of Internet Service Providers (ISPs). All of this contributed to a significant decline in the public's confidence in online security. In some cases, real world costs were considerable, especially where repeated website defacements made commercial transactions over the Internet virtually impossible. An organisation called Middle East Virtual Community tried to function as a mediator between the warring factions, but its efforts were largely fruitless. During 2002 the conflict abated, but spectacular individual hacks were and are still seen from time to time. It seems likely that Israeli and Palestinian hackers will continue to react to real world events into the future, and each intensification of physical violence will be accompanied by a comparable increase in cyberattacks.

6. Implications of Interfada

The significance of the Israeli-Palestinian cyberwar does not reside in its long-term or real world effects, all of which remain relatively minor. Rather, the cyberwar is important because many informed observers have seen it as a model for future Internet cyberconflicts, and from it invaluable lessons can be learned. These lessons include a better understanding of the dynamics of cyberwar and the means to prepare for the "inevitable cyber component of future conflicts."[34] The conflict faithfully followed three basic principles of cyberwar. First, cyberattacks immediately followed physical attacks. Second, as time passed the attacks became more numerous, more sophisticated and better coordinated. Third, attackers were especially attracted to high-profile targets like networks, servers and routers in order to maximise the symbolic, political and tactical effects of the attacks and to gain maximum visibility. Furthermore, the struggle involved all of the general characteristics of conventional combat as described in classic strategic theory. An initial period of surprise attacks was followed by a longer phase of adaptation and recovery on both sides. Then the fighting rapidly escalated and brought in third parties, with volunteers from all over the world jumping in. As a result, the pace of cyberarms development, proliferation and tactical innovation increased.[35] It is extremely hard to point out who were the perpetrators and who were the victims in the endless waves of attacks and counterattacks of the Interfada. The conflict, however, highlights the fact that because of unavoidable collateral damage, ordinary citizens and businesses tend to suffer the most in a cyberwar. The online community as a whole was on the receiving end when viruses, worms and Trojan horses spread beyond their intended targets and infected systems around the globe.

The conflict can also be seen as a symptom of a more general trend in the contemporary Muslim world: the emergence of the Internet as an increasingly important source of information and communications tools. Dr. Limor Yagil, an expert on online terrorism at Tel Aviv University, stated, "The Muslim world understood the importance of the Internet very early. They adopted a new strategy of online Jihad or e-Jihad. They created an Islamic community on the Internet." According to Gary R. Bunt, a leading expert on Islam and the Internet, this new Muslim media culture has brought about two radical new concepts. The first is the Electronic Jihad ranging from aggressive hacking and online militancy to the peaceful coordination of protests. The second is the presence of religious authority on the Internet, in particular "Cyber Islamic Environments," or websites dedicated to the promotion and exercise of Islam.[36]

Although Islamic websites are constantly increasing, we need to keep the broader context in perspective. There is a huge technological gap in the Middle East separating Israel from the Arab population. Technological illiteracy is a major problem among people in the West Bank and Gaza strip, where only 11 percent of households have computers. Most importantly, a mere 3.6 percent of Palestinians - 145,000 people - use the Internet, while in Israel the number is 43.5 percent: over three million users. Although Palestinian Internet usage has in fact tripled since 2000, largely as a result of the establishment of Internet cafés by charity organisations like Enlighten and Across Borders, the overall picture remains the same. Israel is a true world power in the Internet economy, home to many of the most important Internet companies - especially in the field of computer security - a global leader in research and development spending, and second only to the United States in pioneering new Web technologies. Because the Israeli government has forbidden Palestinian ISPs from using direct connection to the Internet backbone, they have to lease their bandwidth from Israeli hosts. Vulnerability resulting from this was clearly demonstrated on 15 July 2002, when the IDF stormed the offices of Palnet, the leading Palestinian ISP, arrested six employees and shut down all of its connections for a day. Thus the situation in cyberwar is not at all dissimilar from the physical world conflict, where the Palestinian insurgents are driven by the fact that they have so little to lose.[37]

The undeniable connection between the intensification of physical violence and the proliferation of cyberattacks deserves attention. All recent major political, social and military conflicts have quickly spilled over to cyberspace, and the fluctuations in cyber activities have faithfully reflected the changes in global political tensions. Although the direction of influence is usually from the physical world to cyberspace, there is also considerable evidence of the reverse. The Internet has prompted and encouraged physical world political action by creating communities of the likeminded and providing them with channels of communication. The anti-globalisation movement, the international campaign to ban landmines, and the Chinese Falun Gong movement are examples of real world activities spurred by online actions.[38]

Could a similar phenomenon also take place in a cyberwar?

Cyberattacks are perpetually increasing in volume, coordination and sophistication. Could online hostilities generate physical aggression by more traditional means, as claimed by some hardcore proponents of restricting freedom of speech on the Internet? This seems highly improbable to say the least. Although there are some accounts of online aggressions by governments, the overlapping of principal actors in physical war and cyberconflicts is almost zero, while online and physical varieties of social activism are predominantly practiced by the same people. Social activism spilling out of online actions into the physical world represents merely an extension of the battleground. Cyberwar would require both significant vertical and horizontal escalation of the struggle. This accounts for the fact that all cyberwars thus far have only been reactions to and repercussions of what has happened in the physical realm. One could even see cyberconflicts in a positive light, as Arquilla and Ronfeldt did in 1993:

> It is hard to think of any kind of warfare as humane, but a fully articulated cyberwar doctrine might allow the development of a capability to use force not only in ways that minimize the costs to oneself, but which also allow victory to be achieved without the need to maximize the destruction of the enemy. If for no other reason, this potential of cyberwar to lessen war's cruelty demands its careful study and elaboration.[39]

In fact it would be beneficial if terrorist groups would start to rely more on cyberattacks instead of bombing, for it would mean fewer human casualties. This is also why the most important considerations in cyberwars are in fact not military but those related to civil liberties, freedom of speech and utilizing the full emancipatory power of the Internet.

7. Freedom of Speech on the Internet

According to Reporters Without Borders, 45 governments currently restrict their citizens' access to the Internet, ranging from all-out denial of use to forced subscriptions to a state-run ISP.[40] Several governments have put forth proposals to restrict the use of cryptography so that terrorists could be more easily revealed, but there is more at stake. Encryption tools like PGP (Pretty Good Privacy) have, for example, allowed dissidents and human rights activists to protect their communications from repressive governments. One of the harshest governments when it comes to cracking down on online activities is

China. Beijing has introduced broad controls on the use of the Internet, banned thousands of websites for political reasons, closed hundreds of unauthorized Internet cafés, arrested dozens of suspected hacktivists, and even ordered the hacking of sites belonging to dissident groups.[41] Following a demand by the Chinese government, Microsoft agreed to censor the Chinese version of its blog tool, MSN Spaces, by making it automatically reject certain words like "democracy" and "Dalai Lama"; and more recently Google, Cisco and Yahoo have joined Microsoft in accepting such an arrangement in China.[42] This shows that Chinese developments can in fact affect the whole global online society.

Most of the technological tools used by militants and activists are either commercial off-the-shelf products or downloadable freeware and shareware programs available to anyone. Governmental efforts to limit access to them would be not only extremely unlikely to succeed but also very questionable from the point of view of civil liberties. Technologically, it would truly mean turning back the clock. The only way to prevent terrorists from communicating through the Internet would be to un-invent it. As long as their websites do not contain any outright criminal material inciting violence or promoting racial hatred, for example, ousting militant groups from the Web would contradict ideals of free speech. Compromising these ideals would be too high a price to pay for silencing extremist voices, even more so when it has been estimated that child pornography sites and various racist and xenophobic sites together represent only about three percent of the total content of the Internet.[43] In fact the most functional method from the point of view of surveillance and eventual criminal punishment would be to leave these sites undisturbed and monitor them to gather information.

The Internet is one of the most complex and demanding judicial challenges that we face. Even the most fundamental issues like privacy, censorship, encryption, e-commerce, intellectual property rights, governance and online civil liberties are still open, and chances of finding solutions acceptable to all parties involved are slim. This is of course also linked to the countering of online terrorism, crime and vandalism. The Internet is a grey area of international and domestic legislation, where traditional instruments of state power are either inadequate or inappropriate. There are very difficult judicial decisions to be made, and it is unclear by whom and how these decisions could be reached in a way that would guarantee international compliance.

Does the Internet need policing and if so, by whom? Who will provide security on the Web, how will the perpetrators of cyberviolations be identified, and what are the proper legal responses? Most of these questions come down to choices between privacy and security. We should somehow be able to simultaneously guard our privacy against the overzealous governmental and corporate attempts to increase electronic surveillance, protect critical infrastructures from cyberattacks, and stop the widening of the "digital divide" between those who have the access to and knowledge of new technologies and those who do not.

Any legal, technical or military response to cyberviolations must take into account possible collateral damage. The introduction of any regulatory mechanisms unavoidably contains a risk of infringements of the privacy of ordinary citizens. In the networked world of the Web, filled with overlapping interconnections, the risk of hitting innocent bystanders is obvious. The Internet is simply too vast, too powerful and too unpredictable to control. Even the strongest actors, such as states, can at best control less than ten percent of the electronic infrastructure they use. Traditional state surveillance is impossible, for without knowing precisely who and what to look for it would require screening for every possible variation. The standard option of forcing a site to shut down or even swarming the target with denial of service attacks would only bring a temporary and limited victory, because the site could easily reopen somewhere else.[44]

In the 1999 Kosovo conflict, NATO commanders deliberately refrained from bombing ISPs or shutting down their satellite links, although they at the same time heavily targeted the facilities of Serb media. Full and open access to the Internet was seen to foster democratic development and to help unveil the crimes of the Milosevic regime.[45] Reid Fleming, an activist from a high-profile hacker group Cult of the Dead Cow, argued that a nation's free access to information would certainly not be improved by attacking its data networks. Cyberattacks are detrimental to freedom of speech. Swarming, flooding and e-mail bombing make normal operation of target sites impossible, amounting to a form of online censorship.

The online world of the Internet represents a new public space. Instead of introducing questionable legal sanctions to control it, states should extend civil rights into the virtual world as soon as possible. The Internet is a revolutionary low-cost, uncensored mass communication medium, a forum for unrestricted social and political debate invaluable to millions of ordinary people all over the world. It is essential to safeguard its freedom from both malicious hackers wanting to misuse it as a tool for inflicting harm on other people and governments overeager to control its use under the pretext of protecting their citizens.

Mark M. Pollitt stated in 1997 that "computers do not, at present, control sufficient physical processes, without human intervention, to pose a significant risk of terrorism in the classic sense." Although many things have changed since then because of rapid and continuous development of IT and especially as a result of 9/11, all of his main conclusions still hold true. Although IT dependency puts information at risk, this risk does not carry any direct threat to our physical safety. This must not, however, lead to carelessness in protecting the information infrastructure. The only way to prevent this infrastructure from becoming a more serious vulnerability is to ensure sufficient human oversight and intervention into technological processes, and to continue developing corresponding defensive measures.[46]

Notes

[1] "Monitor: Interfada," *Information Warfare Monitor* (20 July 2005). <http://www.infowar-monitor.net/modules.php?op=modload&name= Archive&file=index&req=listarticles&secid=5>; Patrick D. Allen and Chris Demchak, "The Palestinian-Israeli Cyberwar," *Military Review* 53 (2003): 52-59.

[2] Allen and Demchak, 52-55.

[3] John Arquilla and David Rondfeldt, *The Advent of Netwar* (Santa Monica: RAND, 1996), 17-24; Dorothy E. Denning, "Activism, Hacktivism, and Cyberterrorism: The Internet as a Tool for Influencing Foreign Policy," in *Networks and Netwars: The Future of Terror, Crime, and Militancy*, eds. John Arquilla and David Ronfeldt (Santa Monica: RAND, 2001), 242.

[4] Julien Pain, "Let's Not Forget 10 September 2001," *Reporters Without Borders*, 22 June 2004 (15 September 2004). <http://www.rsf.org/article.php3?id_article=10760>.

[5] David Ronfeldt, *Tribes, Institutions, Markets, Networks: A Framework about Societal Evolution* (Santa Monica: RAND, 1996), 2-4; Karen Stephenson, "What Knowledge Tears Apart, Networks Make Whole," *Internal Communication Focus* 36 (1998): 22-25.

[6] John Arquilla and David Ronfeldt, "Cyberwar Is Coming!" in *In Athena's Camp: Preparing for Conflict in the Information Age*, eds. John Arquilla and David Ronfeldt (Santa Monica: Rand, 1997), 25-27. In addition to Arquilla and Ronfeldt, Manuel Castells's *The Power of Identity* (1997) and Chris Hables Gray's *Postmodern War* (1997) make important contributions to the theorizing of the non-military end of the conflict spectrum.

[7] Robert J. Bunker, "Battlespace Dynamics, Information Warfare to Netwar, and Bond-Relationship Targeting," in *Non-State Threats and*

Future Wars, ed. Robert J. Bunker (London: Frank Cass, 2002), 101-102; Arquilla and Ronfeldt (1997), 23-24.

[8] Bunker, 100-101; John Arquilla and David Ronfeldt, "Afterword (September 2001): The Sharpening Fight for the Future," in Arquilla and Ronfeldt (2001), 363-364.

[9] Michele Zanini, "Middle Eastern Terrorism and Netwar," *Studies in Conflict and Terrorism* 22 (1999): 248; Bunker, 103-105; Arquilla and Ronfeldt (1997), 49-50.

[10] David Ronfeldt, "Netwar across the Spectrum of Conflict: An Introductory Comment," *Studies in Conflict & Terrorism* 22 (1999): 190; Zanini, 248-251; Giles Trendle, "Cyberwars: The Coming Arab E-Jihad," *Middle East* 322 (2002): 5.

[11] Daniel Verton, "New Cyberterror threatens AF," *Federal Computer Week*, 3 May 1999 (12 July 2004).
<http://www.fcw.com/fcw/articles/1999/FCW_050399_431.asp>; Athina Karatzogianni, "The Politics of 'Cyberconflict,'" *Journal of Politics* 24 (2004): 49; Arquilla and Ronfeldt (1997), 27.

[12] Arquilla and Ronfeldt (1997), 30-32, 40, 45.

[13] Denning, 241.

[14] "Population Explosion," *ClickZ Network*, 3 November 2005 (26 March 2006).
<http://www.clickz.com/stats/sectors/geographics/article.php/5911_15115 1>.

[15] Denning, 239-240; Ashley Dunn, "Crisis in Yugoslavia - Battle Spilling Over onto the Internet," *Los Angeles Times*, 3 April 1999, sec. A, p. 10.

[16] Michelle Delio, "It's (Cyber) War: China vs. United States," *Wired News*, 30 April 2001 (5 May 2004).
<http://www.wired.com/news/politics/0,1283,43437,00.html?tw=wn_story _related>; Karatzogianni, 51; Alex Perry and Nelly Sindayen, "Getting Out the Message," *Time Europe*, 6 April 2001, 84-86.

[17] Allen and Demchak, 53.

[18] William Elison, "Netwar: Studying Rebels on the Internet," *Social Studies* 91 (2000): 127-131; Bunker, 100; Ronfeldt (1999), 190.

[19] Rondfeldt (1999), 190-191; Denning, 281-284; Karatzogianni, 47-48.

[20] Kevin Anderson, "United States Names Cyber-terrorism Czar," *BBC News Online*, 10 October 2001 (7 May 2004).
<http://news.bbc.co.uk/2/hi/americas/1590398.stm>.

[21] James Adams, "The Future of War," *Interactive Week*, 27 November 2000, 52-53.

[22] David B. Ottaway, "US Considers Slugging It Out with International Terrorism," *Washington Post*, 17 October 1996, sec. A, p. 25; Zanini, 250-

251.

[23] Perry and Sindayen, 84-85.

[24] U.S. National Infrastructure Protection Center, *Cyber Protests: The Threat to the U.S. Information Infrastructure*, October 2001 (20 July 2005), 6. <http://www.mipt.org/pdf/cyberprotestsnipc102001.pdf>.

[25] Mark M. Pollitt, "Cyberterrorism: Fact or Fancy?" *Proceedings of the 20th National Information Systems Security Conference*, October 1997, 285-289.

[26] Michele Zanini and Sean J.A. Edwards, "The Networking of Terror in The Information Age," in Arquilla and Ronfeldt (2001), 46-47.

[27] Ibid.

[28] Opening words of the acclaimed official study by the National Research Council, *Computers At Risk: Safe Computing in the Information Age* (Washington, DC: National Academy Press, 1991).

[29] Lee Hockstader, "Pings and E-Arrows Fly in Mideast Cyber-War," *Washington Post*, 27 October 2000, sec. A, p. 1.

[30] Robert Lemos, "'Hacktivism': Mideast Cyberwar Heats Up," *ZDNet News*, 5 November 2000 (30 March 2004). <http://news.zdnet.com/2100-9595_22-525308.html?legacy=zdnn>.

[31] Carmen J. Gentile, "Hacker War Rages in Holy Land," *Wired News*, 8 November 2000 (5 April 2004). <http://www.wired.com/news/politics/0,1283,40030,00.html>; Allen and Demchak, 53.

[32] Brian Krebs, "Hackers Worldwide Fan Flames in Middle East Conflict," *ComputerUser.com*, 25 November 2000 (20 January 2004). <http://www.computeruser.com/news/00/11/25/news1.html>; Carmen J. Gentile, "Palestinian Crackers Share Bugs," *Wired News*, 2 December 2000 (20 January 2004). <http://www.wired.com/news/politics/0,1283,40449,00.html>; Gentile, "Hacker War Rages in Holy Land"; Allen and Demchak, 53-57.

[33] Delio, "It's (Cyber) War: China vs. US"; Allen and Demchak, 53-57.

[34] Allen and Demchak, 52.

[35] Allen and Demchak, 54-57.

[36] Gary R. Bunt, *Islam in the Digital Age: E-Jihad, Online Fatwas and Cyber Islamic Environments* (London: Pluto Press, 2003); Tania Hershman, "Israel Discusses the 'Inter-fada,'" *Wired News*, 21 January 2001 (20 July 2005). <http://wired-vig.wired.com/news/politics/0,1283,41154,00.html?tw=wn_story_related>.

[37] "Internet Usage in the Middle East," *Internet World Stats*, 24 March 2005 (9 July 2005). <http://www.internetworldstats.com/stats5.htm>; Jessica Steinberg, "Facets of the Israeli Economy - The Communications

Industry," *Israeli Ministry of Foreign Affairs*, 2 January 2002 (15 January 2004).
<http://www.mfa.gov.il/MFA/MFAArchive/2000_2009/2002/1/Facets%2 0of%20the%20Israeli%20Economy-%20The%20Communications>.
[38] Michael A. Vatis, "Cyberattacks during the War on Terrorism: A Predictive Analysis," *Institute For Security Technology Studies at Dartmouth College*, 22 September 2001 (20 July 2005). <http://www.ists.dartmouth.edu/analysis/cyber_a1.pdf>;   Karatzogianni, 47, 52.
[39] Arquilla and Ronfeldt (1997), 45. The first version of the article was published in 1993.
[40] "Internet Censorship: Law and Policy around the World," *Electronic Frontiers Australia*, 28 March 2002 (15 April 2004). <http://www.efa.org.au/Issues/Censor/cens3.html>.
[41] Karatzogianni, 51; Denning, 277.
[42] "Microsoft Censors Its Blog Tool," *Reporters Without Borders*, 14 June 2005 (16 June 2005). <http://www.rsf.org/article.php3?id_article=14069>. Tom Zeller, "Web Firms Questioned on Dealings with China," *New York Times*, 16 February 2006, sec. C, p. 1.
[43] Robert Ménard, "A call for vigilance," *Reporters Without Borders,* 2004 (5 March 2005). <http://www.rsf.org/rubrique.php3?id_rubrique=433>.
[44] Bunker, 106.
[45] Denning, 240.
[46] Pollitt, 289.

The New Minutemen:
Civil Society, the Military and Cyberspace

Mark L. Perry

Abstract: Since World War I we have witnessed the accelerated blurring or elimination of the boundary between combatants and civilians. The conduct and aftermath of war now have unlimited effects on society at large. With the rise of the Internet and civil society's increasing sophistication in the use of communications media, this process has only accelerated, as seen in the enthusiasm with which the film "Fahrenheit 9/ll" was greeted in the United States. That war directly affects civilian life on a global basis is becoming clearer. But as yet little understood are the hidden ways in which daily life has been organized on the basis of military culture. Our primary social institutions - schools, hospitals, corporations, governments - are heavily influenced by paradigms created and perpetuated by the military. Several significant anti-war movements have occurred in the last half century. Are the influences of war and of military culture declining and a new influence originating in civil society rising? If we accept the premise that civil society will increasingly influence the military's functioning, then it is possible to argue that the blurring of the line between the soldier and the civilian means not only that civilians are vulnerable, but also that they are effective in influencing the causes, conduct and outcome of warfare. By gaining greater access to global information they are able - whether or not governments approve - to play a greater role in decision-making. If power hinges on information, then as information becomes ever more accessible, power becomes ever more diffused. This new factor therefore counterbalances the centralization of the technologies of warfare, and challenges us to redefine the meaning of power.

Keywords: Internet, Abu Ghraib, Iraq War, Civil Society, Globalisation, Noosphere

1.   Introduction

The so-called Minutemen at Lexington and Concord in the opening stages of the American Revolution responded with famed alacrity to the call of defence against the British forces. But that was not their only contribution to the eventual American victory. The British were bound to a traditional fighting strategy based on strict regimentation, the preservation of ranks, and the firing of weapons according to a formal procedure derived from the discipline of Roman soldiers and modernized by the Dutch beginning in the late sixteenth century. The Minutemen, however, were having none of that. Theirs was a guerrilla strategy of shooting on the run and making use of natural and man-made features for protection, thereby ignoring and successfully overcoming, at least at that initial stage in the war, the British technique. And although throughout most of the war the American colonists relied primarily on such traditional techniques, it is evident that the Minutemen's abandonment of tradition was a harbinger of profound changes to come.

Today the civil society movement's position against Western corporate dominance of the globalisation process is analogous to that of the American colonists. It is materially and numerically the underdog, appearing to lack the resources to achieve its goals. And yet it has demonstrated notable effectiveness, beginning with the mass protest at the 1999 World Trade Organisation meeting in Seattle. The Seattle demonstrations halted the WTO meeting; achieved unprecedented international publicisation of the issues they advocated, establishing in the process a number of alternative news media (most notably indymedia.com); established a long-term and possibly permanent alliance between interest groups that had previously been quite disparate, particularly labour and environmental activists; achieved, through the agency of the Internet, an unexpected and growing degree of social influence, regionally and internationally; and enabled the civil society movement to attain both public recognition and, perhaps more importantly, a firm and unifying sense of self-consciousness within the great diversity represented by its constituent groups.

Thus the civil society movement has created a non-violent variant on classic guerrilla strategy: rather than relying on the large and cumbersome mainstream institutions and organisations created and dominated by its adversaries, it has created its own means and methods that emphasize the role of independent and highly mobile individuals and groups. Like a swarm of bees, it has been highly effective in turning aside conventional organisations, no matter how powerful. It has influenced and altered the course of recent economic and political developments. What had been envisioned as a straightforward march of the West, following its defeat of communism, toward a future of markets dominated by corporations has become disrupted by debate and powerfully-voiced dissension. The top-down system of military culture, which for centuries the West had taken for granted, was first seriously challenged by the civil rights, Hippie and anti-war movements of the 1960s. Now the civil society movement, armed with the latest communications technologies, has redefined "grassroots" in a way that may compel a reformation of military culture, and may reduce our society's emphasis on top-down hierarchy in general.

As one observer has noted, whereas World War II was monitored by radio and the Vietnam War by television, the Iraq War is the first to be monitored by Internet.[1] If knowledge is power, it is clear that the new communications media have supplied the grassroots with unprecedented power by "turning consumers of media content into producers of media content."[2] This phenomenon has most recently been demonstrated by the increasing influence on mainstream media, for good or ill, of bloggers.

The corollary of knowledge as power is that power can also be gained through the control of secrets. This is where the new media transforms the old system; the sources of knowledge and its dissemination have traditionally been in the hands of the elite. Now the global masses are able to bypass the elite's resources and interact directly with each other. The knowledge hierarchy, therefore, is radically subverted. The new paradigm argues "the principle of media as a commons, not a commodity."[3]

In late May 2004 US Defence Secretary Donald Rumsfeld issued an order in an attempt to regain control of information coming from American troops serving in Iraq. "Digital cameras, camcorders and mobile phones with cameras have been prohibited in military compounds in Iraq." Moreover, a "total ban throughout the US military" was apparently being prepared.[4] Such a ban may or may not prove effective in the military, as censorship has met with varying degrees of success in previous wars. But the wider society would not likely accept the application of such a ban in civilian life.

In any case, it appears that the new communications technology has forced a reassessment of military practice, and has begun to shake the paradigm of top-down hierarchy in society as a whole and to increase the voice and authority of the grassroots.

2.   A Revision of Orwell

In George Orwell's *1984* society is controlled by a military culture that has reached the pinnacle of unconstrained social power via high technology. It was and is a dark vision. Yet little did Orwell - or anyone else, for that matter - imagine that the very same communications technology designed for top-down social control would evolve into a two-way conduit of power. We can call this Orwell's Millennial Surprise.

Before pursuing this optimistic view we must acknowledge that Orwell was at least half right and astonishingly so. In 1996 President Bill Clinton signed the Telecommunications Reform Bill that allowed for the free trade of communications companies. Rapid consolidation of media outlets followed thereafter. Today the world's media companies controlling virtually all the major sources of information have been reduced to six.[5] As privacy and freedom of information and ideas have steadily declined, especially since 9/11, private and governmental security forces have greatly expanded. Private security companies are amongst the fastest growing sectors of the service economy.[6] Intimately related to the proliferation of private security are the steady reduction of public space in urban areas and the transfer of commerce and, in some cases even government offices, to privately held property, namely malls and gated communities. Even in the remaining public areas closed circuit security television is becoming standard and even ubiquitous. Other surveillance technologies abound and are diversifying rapidly, each new generation of equipment taking advantage of the regular exponential increase in computing power.

What Orwell did not anticipate, however, was a counter-movement, rising slowly and fitfully from the humblest grassroots. "For more than fifty years, America has nurtured a vibrant, alternative, overtly political media voice."[7] Following World War II, as the consolidation of corporations began in earnest, the Beats, the Hippies, the civil rights and women's movements all fostered a fertile environment for so-called "underground" literature in the form of pamphlets; newsletters; regularly produced news periodicals; literary, political and cultural journals; even simple bulletin boards and ingenious "switchboard" services that disseminated useful information to local communities. Rock-and-Roll and hip-hop music could also be interpreted as forms of underground or alternative communication hidden in plain sight. And just as the youthful social movements seemed to have breathed their last as the disco era arose, the roots of the Internet were sown, opening the field to a fruitful harvest of transcendent media power that seems tailor-made for the decentralisation of knowledge and knowledge-power.

It is, of course, possible that the Internet, like the various forms of popular music mentioned above, will fall under the control of commercial and political interests. The government of China has in recent years attempted to harness the Internet to its purposes, or at least to prevent the use of the Internet against its interests. Yet according to a 2002 study, China has not been able to achieve anything more than mixed success, and even this, the author suggests, will most likely be short-lived. A 2004 study found that the Russian government is in fact gaining significant control over Internet usage in its society and will likely continue to consolidate that control in the foreseeable future.[8] What we see, then, is that the battle lines are being drawn in cyberspace. This gathering storm has been represented in many artistic works, from the recent Matrix film trilogy to the groundbreaking and prescient cyberpunk fiction of William

Gibson on which the films were based, work that - surprisingly - was first published as early as 1984.

Is the cat of genuine participatory democracy well and truly out of the bag, or is our post-industrial, post-modern society doomed to remain only an illusion of decentralisation, local control and individual initiative? It is not possible to say with any certainty so early in the age of globalisation. However, if advocates of democracy have any cause for optimism it is in the fact that the quest for democracy, if not its actual achievement, has become more fervent and self-conscious on a global scale than ever before. While to some it may be a debate between calling the glass half full or half empty, one thing is clear: the new and unprecedented consciousness of grassroots global interaction and mutual identity has continued to grow in strength from the beat generation to our time, and with no end in sight. If the optimists are right, the very fact that state and commercial powers seek to control the Internet is proof that, like jazz and Rock and Roll, the new mode of grassroots e-interaction is here to stay.

3.   A New Social Order

It is not simply the technology of the new media alone that is important. There has also occurred a subtle change in our social relations, a change inherited from the youth movements of the 1950s and 1960s.

In the 1999 "Battle in Seattle," the world witnessed for the first time a large and extremely diverse group of protestors who had organized themselves by means of the Internet and without any traditional hierarchical institution. The Internet, email, cell phones and personal digital assistants (PDAs) tend to foster horizontal rather than vertical ordering.[9] To be sure, the Direct Action Network was a central group that coordinated much of the pre-protest preparation, including basic rules, housing and meeting spaces; but its relationship to the variety of protesters was as a facilitator, and only rarely as a decision-maker. As a result the protest's organisation was quite real yet in a completely different form than American society had ever seen. It was a fluid form, making use of sophisticated communications media to respond rapidly to changing circumstances in highly inventive and creative ways.[10]

Together these two innovations - Internet-based communications and horizontal networking - are creating a new culture that redefines our relationship to hierarchy.

4.   Convergence of Battlefield and Town Square

Since Seattle the civil society movement has spread around the world. The commencement of the war in Iraq was opposed by what was essentially a Seattle-style protest on a global scale, some 10 million strong. Although the demonstrators were unable to stop the war they were able to attain a significant goal: they gained a voice, one powerful enough to be heard on an international level, truly a shot heard round the world. While many observers have correctly observed that such protests are composed of a seemingly chaotic diversity of interests, it does not necessarily follow that this diversity spells internal conflict and the dilution of their respective messages. As one Seattle participant has noted, it was precisely because sea turtle advocates and longshoremen joined forces that a transcendent, constructive social movement was created.[11] Forced to find common ground, the various special interest groups were able to create a framework that enabled any and all constructive efforts to unite. The Battle in Seattle established a new paradigm: social action is more far-reaching and profound if it embraces the principle of unity in diversity rather than unity in uniformity.

Unity in diversity inevitably implies the unity between the protesters and the mainstream society itself. We have been witnessing over the past century the strengthening of economic, transportation and communications ties between the nations of the world, such that we can now refer to our planet as a "global village." Social change is rapid and accelerating. Even the most distant villages in the developing world, for good reasons and bad, are quickly becoming included in the world economy and world politics. Very few spots on the globe are still isolated, as eco-tourists would sadly testify. We can therefore recognize that any perceived distance between the town and war, physical or metaphorical, is no more. We began to understand this in World War I, when for the first time the world experienced what is now known as total war, a blurring or elimination of the boundary between combatants and civilians on a global scale. Since then this process has only accelerated. And here we can easily recall the strange spectre of the most remote South Pacific island paradises transformed into ghostly nightmares in World War II.

Usually in this context we think of the civilians as being caught up on the battlefield. Now, however, we should recognize that the converse is also necessarily true: the battlefield is being caught up in the actions of civilians in the town square. Perhaps the most dramatic evidence of this is the unprecedented enthusiasm with which Michael Moore's documentary *Fahrenheit 9/11* was greeted in the United States and internationally.

How is that possible? The same transportation and communications media that make possible the conduct of modern warfare also enable civilians to inform themselves about a war, comment on it, and even influence decisions about it. This first became obvious in the course of the Vietnam War, in which the images broadcast on television fuelled protests that were ultimately successful in removing American troops from the field. With the advent of cyberspace, the flow of information to citizens is now approaching constant real time, and their participation in town square debates is all the better informed, their voices all the more authoritative and powerful. It is for this reason that Michael Moore was able to observe in *Fahrenheit 9/11* that the opposition to the Iraq War was formed far more quickly than the opposition to the Vietnam War. With the advent of digital technology, moreover, the voices of civilians and journalists can be more powerful than those of the military leadership simply because civilians might have more knowledge.[12] For what military, governmental or media institution can match the information-gathering power and speed of a vast network of privately owned and operated digital recording devices and laptop computers?

Cyberspace, therefore, has become the new town square in the global village. More importantly, because cyberspace is everywhere, citizenship can also be transnational, independent of time, place or accident of birth. As Jeremy Rifkin has argued, the new media technologies are not about products but about relationships; they build relationships between individuals, groups and nations at lightning speed, and each new generation of equipment only strengthens those bonds.[13] There has been a tremendous synergy between the Internet, the satellite news networks, and personal publishing and communications media. Even though television news companies are owned by corporations often opposed to the civil society movement, they are caught up in the new web. And this worldwide web, which seemed to be such a flimsy tissue, is rapidly becoming a solid and permanent infrastructure transforming the very basis of human interaction and human civilisation. Internet use is rapidly expanding and intensifying: One report found that between March 2000 and June 2004 the American audience for Internet news rose by 70 percent from 54 million to 92 million. Furthermore, "Americans are turning to the Internet for news coverage they cannot find in the mainstream media," a phenomenon that resonated with Michael Moore's film.[14] The photographs and video footage from Abu Ghraib are vivid examples of how personal media technologies have unsurpassed power to set the news agenda, to redefine what is and is not content, independently of the older media companies.

The picture is complicated further by the fact that many different political groups are employing the same digital media, often in a direct tit-for-tat exchange to "win the hearts and minds" of the viewing public, and the highly valued favour of "international opinion." Following the revelation of the Abu Ghraib scandal the world witnessed the beheadings

of Western hostages in reprisal. Information, policies, warnings, threats
and actions are now easily translated from the realm of words and deeds in
the shadows to what might be termed a new kind of realism in the full
presence of the international community.

But for the sake of clarity, let us focus on the role of the Abu
Ghraib digital photographs, which is only gradually becoming clear. In
May 2004, days after the scandal broke in the American news media,
commentators were already arguing forcefully about the centrality of those
images not only to the scandal itself but to the war and the future of
society in the era of globalisation. A.D. Coleman, a photography critic and
historian, declared, "This is as far as I know the first instance where
digitally generated images made by an amateur photographer have erupted
onto the scene of current events and had an impact. But it won't be the
last."[15] Susan Sontag addressed the role of these images in exposing the
critical issue of prior knowledge:

> The pictures will not go away. That is the nature of the
> digital world in which we live. Indeed, it seems they
> were necessary to get our leaders to acknowledge that
> they had a problem on their hands. After all, the
> conclusions of reports compiled by the International
> Committee of the Red Cross, and other reports by
> journalists and protests by humanitarian organisations
> about the atrocious punishments inflicted on "detainees"
> and "suspected terrorists" in prisons run by the
> American military, first in Afghanistan and later in Iraq,
> have been circulating for more than a year. It seems
> doubtful that such reports were read by President Bush
> or Vice President Dick Cheney or Condoleezza Rice or
> Rumsfeld. Apparently it took the photographs to get
> their attention, when it became clear they could not be
> suppressed; it was the photographs that made all this
> "real" to Bush and his associates. Up to then, there had
> been only words, which are easier to cover up in our age
> of infinite digital self-reproduction and self-
> dissemination, and so much easier to forget.[16]

All of these developments shed a different light on the famous
question, "What did you know and when did you know it?" The answer
tends more and more to be: "Everything and almost immediately." The
question is now often moot because the new media create a context of

perpetual information transparency. The shadows of secrecy are disappearing, and the claim of ignorance is less and less plausible.

5.   Long-Term Implications
A. The Case of Rollback

For the sake of analytical clarity, let us consider the possibility that a military ban against personal publishing media is strengthened and extended to the general public. It is doubtful at this time that such a ban would ever succeed - but what if it did? Certainly it is not so far-fetched in more authoritarian countries, and we should anticipate the possibilities.

In that case we would essentially see the elimination of the global village's town square. Cyberspace would cease to exist as a free public space. It would become, as malls are now, a private or governmentally-controlled space masquerading as public space. Rights of self-expression on the Internet would be severely curtailed and monitored. Orwell's nightmare would come closer to fulfilment. The militarised culture of top-down hierarchy would predominate in all institutions. The civil society movement would weaken and perhaps die. Social relationships would revert from a mix of the horizontal and the vertical to predominantly vertical. Individuals would be isolated relative to each other, and would interact primarily with the governmental-technological institutions above them in the social hierarchy.

B. The Case of Open Cyberspace

If, on the other hand, cyberspace remains open and reflects what Karl Popper aptly termed "open society," then we can anticipate a number of profound changes in our relationships to the military and to governmental hierarchy in general, changes that will reverberate throughout our culture.

An open cyberspace culture, a free global town square, will more or less oblige the military culture to be responsible to questions and concerns voiced by the public. If in the era of total war civilians and soldiers are no longer separate, open cyberspace means that civil society and military decision-makers are likewise no longer separate. The artificial line drawn between them has been removed, and military decisions will increasingly be made in the context of what can be seen as indirect and sometimes even direct democracy.

This does not mean that the military will be eliminated, or that government hierarchy will be flattened by a tendency to horizontal social relations. Rather it means that the exaggerated top-down hierarchy of traditional military culture will be gradually tempered, and eventually reformed to accommodate a new social structure that it had previously ignored: civil society. It is ironic that since at least the Vietnam War, civil society and the military have often seen each other as arch adversaries - and never the twain shall meet. Yet in the era of open cyberspace this mutual hostility cannot survive. Each party will be forced to move closer to the position of the other. Adherents of the civil society movement will be obliged to recognize and accept for themselves some of the responsibilities that had previously been the burden of the military alone. And the military will be obligated to recognize and accept for themselves some of the responsibilities for maintaining civil order - civilisation - that had previously been the burden of the home front alone.

The openness of cyberspace means the generalisation of the town square; that is, the entire world itself now becomes the town square. In the past, wars have been fought in an unknown place somewhere out of sight, off the edge of the map, in a space between civilisations, a space that lacks civilisation and is therefore suitable for barbaric behaviour. The advent of the new media has shed light on these dark interstitial spaces and, for better or worse, forced us to examine them and claim responsibility for them. In the process, we have become aware of what had previously been kept secret. Textbook publishers, film and news editors, artists, speech writers and high school valedictorians had all tended to keep those spaces dark and secret. Cyberspace now enables us to reach down and expose the long hidden sweepings under the rug.

Culturally the change has many deep implications. By long tradition military personnel have regarded themselves as distinct from civilians. Their uniforms, rituals, educational and training institutions, secrets, and even laws and court system set them apart from the civilian population. For their part, the civilians have perhaps exaggerated their own cultural ways, choosing, sometimes consciously as in the 1960s, to reflect the reverse of military life: long hair, unorthodox clothing, slowness and lack of discipline.

It seems that the lesson of Abu Ghraib is that we are all responsible in the end for the prosecution of war and the preservation of peace, that ultimately there is no real distinction between the military and civil society. We have been playing a game of tug-of-war, each side concerned that it will be pulled into the camp of the other and thereby become powerless to attain its goal. Open cyberspace has revealed a new reality, or at least a reality that if not new was long hidden. Ultimately there can be no separation between the functioning of the military and the functioning of civil society.

Hierarchy, then, has been distorted and severely misconstrued. On the one hand it is needed to a certain degree in order to facilitate organisation. The Internet itself is organized according to a hierarchy of routers, servers and users. But what we have done throughout most of history is to interpret hierarchy according to a values system; we have believed that those individuals and institutions in the upper ranks of the hierarchy are inherently superior to those in the lower ranks, and therefore their knowledge, their vision, their being is inherently superior in all circumstances and questions. Systems philosophy shows the flaw in this reasoning. Any complex system is composed of what can be viewed as a hierarchy of subsystems.[17] The human body, for example, is a single organic structure in which are united thousands of sub-structures: organs, vessels, tissues. And then there are the more subtle aspects, energy flows that we have only just begun to understand. In some cases there is indeed a ranking in importance. The brain is more important than the right ring finger, for example. But in many cases it does not make sense for hierarchy to be the basis of understanding. To take a second example, we know that the entire human body can be quickly destroyed by the malfunctioning of an organ as lowly as the pancreas. Each element of the system contributes to the regulation of the whole.

In the light of open cyberspace the exclusively top-down flow of orders from political and economic elites to the civil society is now being reconsidered. As in the functioning of an organic system like the body, there are certainly countless situations in which the higher order elements in the system rightly instruct the lower. But there are also countless conditions in which the lower order elements inform the higher to the benefit of the whole. By strengthening our use of the Internet and our understanding of the global town square created by open cyberspace, we are learning that social hierarchy is, like the Internet itself, a two-way conduit.

By means of the Internet we have created transcendent technological powers, computer-mediated interaction between people that balances hierarchy with the grassroots. In the final analysis, the global town square can emerge only as a space defined by both the vertical and the horizontal.

A pessimist, however, may expect that the new media will, like all earlier forms of technology, become the tools for unprecedented destruction, as we have already seen in the 9/11 and 3/11 attacks. The Internet provides no guarantee that all its fruits will be beneficial. Yet an optimist might hope that the use of such media to establish, on a permanent basis, a global open society would, if not immediately preclude, gradually dissolve any perceived need to express political will by means of violent acts, by terrorism symbolising the desires of those not represented at the global table. For with a voice, and with ears to hear, and the attention of the global grassroots, it is perhaps possible that a new balance could be attained that transcends the traditional antagonism of adversaries. Perhaps we can create transcendent social powers through a new interaction between civil society and governmental and military institutions throughout the world.

6.   War, Civil Society and the Noosphere

Assuming that civil society and governmental agencies - including the military - continue to converge, we can envision more clearly the emergence of what Pierre Teilhard de Chardin (1881-1955) referred to as the noosphere, or planetary consciousness. When the concept was published in his posthumous work *The Phenomenon of Man* (1959) the world could understand little if anything of it. It was the epitome of optimistic philosophising about the potentialities of the human race, but it seemed far removed from the reality of the post-World War II international situation. Indeed, the Cold War inaugurated an era in which a new, unprecedented and profound pessimism swept the intellectual world. Teilhard de Chardin's vision was considered a mystic fantasy, a rich dream, a foolish hope.

Now, a short half-century later, it is difficult for us to understand how unrealistic his vision seemed. We have quickly become accustomed to the globalisation of thought through the Internet. What has made the difference, and how will this affect the relationship between civil society and war?

A primary cause of our change of perspective is, to put it simply, a matter of coordination and speed. For millennia there has been long-distance communication between peoples, even across continents and oceans. What characterizes the age of the noosphere is that this communication suddenly has acquired the capacity to approach the speed at which natural thought occurs. Whereas in the past the coordination of global thought was difficult because of the long delays incurred in the transmission of messages, now the Internet has accelerated the process and looks set soon to reach the ultimate goal: convergence in real-time.

Real-time convergence in this context has a specific definition: the seamless, uninterrupted transmission of text, data, voice and video by one system without delays.[18] But this is only the beginning. The real power of convergence is that it will permit unlimited access to communication in a socially natural way. Since the advent of the telegraph and telephone most technologically-mediated communication has been unnatural: technical circumstances have forced it to be binary, between two parties. But this is not how people naturally communicate. In society communication occurs constantly between an endless variety of groupings: from an isolated individual speaking to herself, to a pair exchanging information, to a small group, to a large crowd listening to a speaker, to vast mass demonstrations as have been seen in numerous "people power" movements. Our communication flows between these infinitely diverse groupings throughout each day. One morning a teacher might begin the day conversing with a single colleague, then in the next hour with a group of five around a conference table, then in the next with a class of forty students, then in the afternoon with two hundred at a faculty meeting. Overlapping all of this may be private and personal communications with family and friends, with local vendors, with travel agents and other service providers. As our communications technology has become more powerful it has enabled us to widen our access to people and groups in real-time. Yet the door to global real-time communications has remained shut except for the few military, government and news media institutions with 24-hour access to satellite links. As a result our consciousness of the global population has been only theoretical, quite indirect, experienced only second- and third-hand via interviews broadcast on television and radio, and via ham radio operators. Most of us have been spectators; we have long known that the global community exists but we have not been able to participate in it directly as individuals. That is until now.

The Internet has for the first time cracked open the door to the global community. It has dramatically increased our access to direct real-time communication in two ways: 1) it has provided real-time and near real-time communication in chat rooms and live telephone and video links; 2) it has made all this available at costs that are affordable to large mass populations East and West, not merely in the developed "North" but increasingly in the developing "South." But this is only the first glimmerings of what is behind the door. We have yet to open the door wide and walk through.

When we do we will find the following: inexpensive or even free access for the masses in the developed and developing worlds to real-time convergence media. In other words, the Internet will metamorphose into a transcendent medium that facilitates natural communication, an infinitely diverse free movement of information between dyads, groups and even masses, a flow of communication that expands and contracts naturally according to the will of the people at any given moment. The closest analogy would be the word-of-mouth communication that is at the heart of small town and village life. It is at once both rapid and potentially accurate, for whatever is misunderstood can be soon corrected by actually contacting "the horse's mouth," the source, which is always nearby.

Here, however, we must emphasize that accuracy is only potential. For just as small town information can be exceedingly accurate, it can also just as easily become twisted and fallacious. The very fact that the town's grapevine is easily accessible entices some members of the community to take irresponsible advantage of it to broadcast rumours, some true and some not. On the global level we have already begun to see this play out on the Internet. Like software viruses, intellectual viruses and memes - falsehoods, deceptions, lies, malicious gossip, and urban legends - abound in certain Internet circles. There is no guarantee that the convergence of communications media in real time will lead to any increase in responsible usage. The question then will be how to separate the wheat from the chaff.

To a great extent the primary responsibility to distinguish truth from falsehood in the noosphere will rest on the individual, just as it does in any local community. But nevertheless the noosphere possesses a quality that characterizes it as a new level in human potential. Whereas in the past the truth was often easily concealed simply because the masses had little or no access to the facts, in the age of the noosphere the suppression of access to facts is difficult if not impossible. This is what we have seen in the Abu Ghraib scandal. The point here is that the noosphere is coming into being in a way that surprises even those who are fully cognizant of the Internet's technological fulfilment of Teilhard de Chardin's vision. For his theory was neither solely nor even primarily concerned with the *technical* means for the emergence of planetary consciousness, but rather the inevitable emergence of the social, psychological and even spiritual unity of consciousness. The significance of the noosphere is that it not only grants access to facts that in the past would have been hidden, but it also represents the creation of a highly persuasive global moral consensus about those facts.

Let us examine this claim more carefully. One of the most important aspects of the Abu Ghraib scandal that is often overlooked, and that is how quickly the consensus was reached that the treatment of the prisoners was wrong on the strength of the information supplied by the personal media. Without such personal media would a moral consensus have been reached so quickly? Might it not have been reached at all?

In the darkness of information blackouts moral confusion is easily achieved, and thus freedom from moral constraints is normal. In the light of the Internet and globalized personal communications, actions are more clearly observed. We all become witnesses, and as witnesses we constitute a disinterested jury. How is this possible, one might ask, given that the world is divided into countless political and ideological communities? This is where Teilhard de Chardin's vision comes into play. It appears that global consensus is possible - despite claims of the inevitability of moral relativism - precisely because human consciousness is fundamentally one. The noosphere exists on the global level because we all accept as true at least the most basic moral values. It has always been only a minority that has opposed the concept of a universal moral creed; but the masses around the world have built their societies on what are essentially common truths, beliefs, rights and responsibilities, codes of conduct that are increasingly becoming consolidated and publicized. A most recent example of this process was The Millennium Peace Summit of Religious and Spiritual Leaders, which preceded The Millennium Summit of the United Nations held in September 2000.

This is not to argue that all matters of legal and moral judgment should be left to the global masses. Certainly cases will always abound in which judgment must be made by means of a meticulous process of investigation and reflection by dedicated professionals. Nevertheless, it is equally certain that in many cases of gross misconduct, global moral consensus will play a unique role in the unfolding of any future international legal order. It is notable that the rise of, or need for, global moral or ideological consensus has been cited by a wide variety of leaders of thought, from Francis Fukuyama to Terry Eagleton.[19]

That the vast majority of people in the world opposed the abuse of prisoners at Abu Ghraib shows that the noosphere is composed of not only an intellectual consciousness but also a moral one. The significance of the Internet, then, is that it does not so much *create* moral consensus as reveal the fact that such consensus already exists.

7.   The Rise of the Non-partisan Gaze

Doubtless this was not the first case in which international moral consensus was achieved. Since at least the formation of the United Nations the "opinion of the international community" has become an increasingly powerful force in influencing or even determining the outcome of international issues. Yet we can readily see that the emergence of cyberspace has greatly magnified the scope, influence and reaction speed of that opinion. In fact, it has become a distinct new power unto itself: it has become transformed from an *opinion* - an intellectual and moral judgment after the fact - to a *gaze*, which has the astonishing power to inhibit criminal behaviour simply by existing in real-time. Jane Jacobs wrote of "eyes on the street" as a primary inhibitor of urban crime.[20] Perhaps the same holds true at the global level. By demonstrating its consciousness of criminal behaviour in real time, the gaze of the world can force potential criminals in the international setting to be likewise self-conscious. As Michel Foucault wrote in describing Jeremy Bentham's panopticon:

> He who is subjected to a field of visibility, and who
> knows it, assumes responsibility for the constraints of
> power; he makes them play spontaneously upon himself;
> he inscribes in himself the power relation in which he
> simultaneously plays both roles; he becomes the
> principle of his own subjection.[21]

But there is a vital difference between the surveillance that Jacobs advocates and that of Bentham's panopticon.

> Safety on the streets by surveillance and mutual policing
> of one another sounds grim, but in real life it is not grim.
> The safety of the street works best, most casually, and
> with least frequent taint of hostility or suspicion
> precisely where people are using and most enjoying the
> city streets voluntarily and are least conscious, normally,
> that they are policing.[22]

The eyes on the street are there not because it is a grim duty commanded by the state; rather they are there voluntarily, representing the free will of civil society in its normal functioning. Thus the gaze is not designed to limit freedom in a totalitarian nightmare; the existence of the gaze emerges from freedom. Civil society is a constructive presence; it inhibits negative behaviour not by the threat of punishment but by the constant presence - at once material and spiritual - of constructive life.

We thus see the international situation changing in three stages: 1) Originally war and other international conflicts were carried out in shadows, darkness and ignorance. 2) Light began to be shed through the establishment of the Geneva Conventions, criminal courts to try war crimes after the fact, and the formation of an increasingly consolidated international opinion on moral right and wrong. 3) International opinion became transformed by real-time global communications into a constant gaze of the ultimate disinterested jury: the human race as a whole.

8.   War, Freedom and the Gaze

This vital point merits further consideration: Have we then lost our freedom? Does the constant gaze of the world community stifle our ability to create and act according to our will? Does it promote Orwell's totalitarian nightmare, or champion human rights simultaneously at the grassroots and global levels?

It is clear that the gaze of the Internet limits freedom in war and in society, and it does so because behind the gaze is a mind, the noosphere, that apparently seeks to condemn behaviour it deems criminal, to build, whether we like it or not, a global moral consensus. In this sense, then, the Internet and the noosphere merely bring to a stage of fulfilment the function of United Nations *observers*. Blue-helmets are at a distinct disadvantage when trying to oversee the conduct of a far-flung war with a handful of lawyers in helicopters and jeeps, wielding binoculars, tape-recorders, video cameras and computers. But their functioning, however limited, has established that observation of war - and indeed of all conflict and crime - is a primary means for preserving human rights, establishing peace, and creating at least a basic moral order.

We have already noted how the Internet expands the scope and power of this observational function. There is, however, a distinct advantage to the gaze of the world via the Internet. Whereas the opinion of a small team of blue-helmets, or any similar group, could be construed as an unreasonable imposition on freedoms - especially if that opinion is in fact wrong - the gaze of the world in the age of cyberspace cannot be taken as a step towards totalitarianism. The totalitarian state, like the administration of Bentham's panopticon, is a relatively small elite using high technology to control the masses in their political and personal lives. By contrast, in the case of the Internet's role in the observation of war and conflict, it is we who are gazing at ourselves. Soldiers and citizens alike are - in a space in which the battlefield and the town square have converged - ensuring the fulfilment of both military and civic duties.

By definition, then, the noosphere and cyberspace represent not the limitation of freedom but rather a constructive self-discipline, which, at least in an optimistic view, ultimately leads to the emergence of hidden constructive potentialities.

Notes

[1] Dean Wright, quoted in Irene McDermott, "Iraq Around the Clock: The First Internet War," *Searcher: The Magazine for Database Professionals* 11 (2003): 10-15.
[2] "Will Mainstream Media Co-Opt Blogs and the Internet?" *World Economic Forum*, 22 January 2004 (17 September 2004). <http://www.weforum.org/site/homepublic.nsf/Content/_S10121>.
[3] Teo Ballvé, "Another Media Is Possible," *NACLA Report on the Americas* 37 (2004): 29-31.
[4] "Rumsfeld Bans Camera Phones in Iraq: Report," *ABC News Online*, 23 May 2004 (17 September 2004).

<www.abc.net.au/news/newsitems/s1114150.htm>.

[5] Time-Warner, Disney, News Corporation (Rupert Murdoch), Bertelsmann, Viacom (formerly CBS), and General Electric (NBC). Ben Bagdikian, *The New Media Monopoly* (Boston: Beacon Press, 2004).

[6] Michael Renner, "Curbing the Proliferation of Small Arms," in *State of the World 1998*, eds. Lester Brown et al. (New York: W.W. Norton, 1998), 131-148.

[7] Makani Themba-Nixon and Nan Rubin, "Speaking for Ourselves: A Movement Led by People of Color Seeks Media Justice - Not Just Media Reform," *The Nation*, 17 November 2003 (17 September 2004). <http://www.thenation.com/doc/20031117/thembanixon >.

[8] Ronald J. Deibert, "Dark Guests and Great Firewalls: The Internet and Chinese Security Policy," *Journal of Social Issues* 58 (2002): 143-159; Marcus Alexander, "The Internet and Democratisation: The Development of Russian Internet Policy," *Demokratizatsiya* 12 (2004): 607-627.

[9] For a description of horizontal ordering see Lawrence M. Friedman, *The Horizontal Society* (New Haven: Yale University Press, 1999).

[10] Alexander Cockburn et al., *Five Days that Shook the World: Seattle and Beyond* (London: Verso, 2000); Janet Thomas, *The Battle in Seattle: The Story Behind and Beyond the WTO Demonstrations* (Golden, CO: Fulcrum, 2000).

[11] Thomas, 17-30.

[12] Ilene Prusher, "In the Midst of the Iraq War - and Always on the Phone," *The Christian Science Monitor*, 27 May 2003 (17 September 2004). < http://www.csmonitor.com/2003/0527/p07s02-woiq.html>.

[13] Jeremy Rifkin, *The Age of Access: The New Culture of Hypercapitalism, Where All of Life is a Paid-for Experience* (New York: J.P. Tarcher/Putnam, 2001).

[14] Deborah Fallows and Lee Rainie, "The Internet as a Unique News Source," *Pew Internet and American Life Project*, 8 July 2004 (17 September 2004). <http://www.pewinternet.org/PPF/r/130/report_display.asp>.

[15] Quoted in Amy Harmon, "With Digital Cameras, the World is Watching," *International Herald Tribune*, 8 May 2004 (17 September 2004). < http://www.iht.com/articles/518970.html>.

[16] Susan Sontag, "Regarding the Torture of Others," *New York Times*, 23 May 2004, Sunday Magazine, 25.

[17] Ervin Laszlo, *Introduction to Systems Philosophy: Toward a New Paradigm of Contemporary Thought* (New York: Gordon and Breach, 1984).

[18] See for example David Greenblatt, *The Call Heard 'Round the World:*

Voice Over Internet Protocol and the Quest for Convergence (New York: American Management Association, 2003).

[19] Francis Fukuyama, *The End of History and the Last Man* (London: Penguin, 1992); Terry Eagleton, *After Theory* (New York: Basic Books, 2004).

[20] Jane Jacobs, *The Death and Life of Great American Cities* (New York: Random House, 1961), p. 35.

[21] Michel Foucault, *Discipline and Punish: The Birth of the Prison* (New York: Vintage, 1979), pp. 202-203.

[22] Jacobs, 36.

Part IV

Parallels

On the Similarities between Business and War

Albrecht M. Fritzsche

Abstract: This paper explores the phenomenon that business and war are often treated in the same way. While this seems quite plausible from a technological point of view, it contradicts the meaning of war and business for human life. In order to resolve the contradiction it is necessary to distinguish between human activity and experience. While experience is always existential, activity can be virtual in the sense of a game which takes place in an artificial environment. Virtual activity does not require any relation to the essence of human life. The historical development of war and business during the last four hundred years has increasingly shifted the treatment of both to the technological framework. Within this framework, they are approached by the same methodological apparatus created by the discipline of Operations Research. Examples from practice show that Operations Research has indeed facilitated a fruitful exchange of methods and ideas from one scenario to the other. Nevertheless, the question of whether one of the two frameworks of thinking can be completely ignored must clearly be answered in the negative. Both frameworks must be kept in mind.

Keywords: Business, War, Operations Research, Game Theory

1.   Similarity of Approaches

Many theories treat war and business in a similar way. In the study of war, a historian will record the course of events. He or she will find out what decisions led to the situation in which the declaration of war took place, and the consequences for the political and social structure of all the countries involved will be discussed. An engineer examining the processes of warfare will analyse the usage of the machinery of war, the particular weapons and other material resources. He or she will also study the continuously growing importance of the role of information and the way it is gathered, transferred and applied. A psychologist will examine, on the one hand, the state of mind that makes participation in war possible and, on the other hand, the effects of war on mental health. He or she will study the influences that make people able and willing to fight, to risk their lives and to harm others, or to reject all of this. A biological approach may even look deeper into the physiological functioning of feelings such as safety, aggression, fear and power.

Most studies of business are very much the same. In the past business history has been neglected by many historians, but lately it is receiving ever greater attention. It records a set of events, decisions and operations which could just as well be military. Likewise the engineer approaches business much as he does war; it makes no difference whether the processes and machinery are used in production and trade or in battle and intelligence. Business psychology is most apparent in advertising, but is just as important in production design, negotiation and human resource management, with questions about obedience, group behaviour and dehumanisation very similar to those raised in fighting. In the case of biology, the similarities between war and business are seen in the parallels between the physiology of safety and saturation; and between stock-piling and greed on the one hand, and fear and aggression on the other.

The perception that war and business are interchangeable is not limited to academic pursuits. Modern thinking employs various "war-as-business" and "business-as-war" metaphors. Language applies specific vocabulary from both scenarios in mixtures. Expressions like "just a job," "does it pay," "bloody defeat," "striking force" can be used all in the same context without raising stylistic concerns. Sun Tzu's *Art of War* is very appealing to managers and aspiring executives. Many other books try to capture this momentum to convince business people that it makes sense to translate aspects of warfare to commerce. An interesting example is Kenneth Allard's *Business as War: Battling for Competitive Advantage*. Allard states that it is not only military strategy and structure that should be adopted in the business world, but also military customs and ethics.[1] At the same time, commercial ways of thinking, such as advertising, compensation and bargaining are not uncommon in contemporary military considerations. Warfare requires huge expenses. Awareness of economic considerations in war can be a key factor for effectiveness as well as for support from the home front industries. It therefore seems as if war and business are basically two aspects of one and the same phenomenon. But is that really true?

2.   Two Levels of Understanding

Trade and combat are some of the most traditional human activities. There is no doubt that they were already flourishing before writing had developed far enough to be perpetuated from one generation to another. In the hieroglyphs of the temples of Carnac we find an early recording of a military battle, probably the siege of Megiddo, known from the Bible as Armageddon. The battle is dated 1468 BCE. At approximately the same time, between 1473 and 1458 BCE, the description of a big trade expedition to the land of Punt on the upper Nile was recorded on the walls of the temple of Hatshepsut. There was no money used in trading between the nations, but the exchange of goods was already quite extensive and brought great wealth to the empire of the Pharaohs. Excavations from various sites around the world reveal weaponry not used for hunting but for killing men, as well as goods which cannot have originated in the areas where they were found but that must have travelled long distances in the bags of traders. War and business have developed at several places around the world independently; and they were natural developments, since any civilisation depends on interaction, both in the military and the economic sense, between people. Differentiated social roles encourage the exchange of goods in a peaceful way just as much as they encourage the use of violence to gain power over others. It seems almost unthinkable that a culture of reasonable complexity could develop without becoming involved in warfare and trade. But war and business are not only shapers of culture; they are at the same time dependant on culture. Each requires institutions, rules, laws and advanced structures of communication. Organised war is unthinkable without calculated behaviour, national or social distinctions, leadership, orders and obedience. Trading is unthinkable without fixed values of material goods, distinctions of interest, shared language and symbols, and bargaining procedures.

The cultural and social inheritance of war and business could not be understood if they were treated like natural phenomena. An approach which applies to evolutionary history, chemical synthesis or a functioning spacecraft is not sufficient for the qualities of war and business. An adequate understanding requires a different framework of study. We can perfectly well understand the functioning of nature and technology without asking questions about their meaning for human existence; to some degree it does not even make sense to ask such questions at all. For the understanding of war and business, however, asking these questions sooner or later becomes inevitable. Is it human to kill other people for no better reason than being ordered to do it? How can a job consume a person's life to such an extent that he or she forgets everything else around them? What is the sense of all this?

3.   The Artistic Framework

Asking such existential questions requires a completely different framework of thinking than the above-mentioned approaches of the historian, engineer, psychologist and biologist. We can no longer distance the object of study from ourselves as spectators. Being involved, we cannot think about statements of abstract logical functioning. The new framework of thinking must refer to our own experience. This can be achieved by an artistic approach, as for instance in literature, painting or music, and in a different way in theology, philosophy and neighbouring lines of thought. A work of art is not understood by rational description, but by personal acceptance in the reception of its essence. Using the words of Bergson, we can say that it places intuition next to rationality as another means of understanding. Each statement is evaluated with reference to an individual experience of human existence and cannot be isolated from it.

It is not possible here to present a complete discussion of the artistic treatment of war and business. The point, however, can be illustrated by discussing two acclaimed works of literature, Leo Tolstoy's *War and Peace* and Margaret Mitchell's *Gone with the Wind*.[2] Tolstoy's novel, first published between 1868 and 1869, depicts the personal destinies of a group of Russians during the Napoleonic wars. The French campaign in Russia is a cleansing process for both the nation and the main characters of the novel. At the end, Pierre Bezukhov finds his happiness with the Rostov family on their farmland. Another, Prince Andrei, who has the intelligent mind of an academic, loses his connection to Russian traditions. He dies, less from a battle wound than from having lost his will to live. On the other hand, Tolstoy does not deny the existence of war as a typical human activity. Wars like the French campaign in Russia are incomprehensible outbursts of violence between cultures. But they are unavoidable, and no one, not even a general like Napoleon, can in any way guide or control them. Wars challenge the existence of each individual involved in them, but Tolstoy makes clear where he believes people can recover: in the Russian countryside. The motif of healthy farming life is also widely present in Tolstoy's other works, for example in *Anna Karenina*. He considered it the backbone of Russian society.

Sixty years after *War and Peace*, Margaret Mitchell's perspective on war and business in *Gone with the Wind* was only slightly different. For Scarlett O'Hara the American Civil War is also incomprehensible and she endures it as she would have endured a natural disaster. She cannot understand the patriotism of the other characters which makes them enthusiastic about the war and drives them to fight until they are completely defeated. Scarlett is neither a pacifist nor a revolutionary; she simply thinks in categories other than pride and military courage. She has a clear sense of the good life and she is willing to work for it. After the defeat of the Confederacy, she accepts the new situation without much questioning and can thus adapt much better to it than those around her. With a clear sense for economy, she establishes a successful business while her old friends still mourn their losses. Just as much as she did not understand why they supported the war, they do not understand how she could possibly behave like a successful businesswoman. But Mitchell makes clear that it is Scarlett's will to live and to rebuild her world by pursuing business that enables her to survive and help the people around her to overcome all their hardships.

As these examples suggest, the artist's view of war and business is focused on their essential meaning for human life. From this perspective there are no similarities between them. In fact, war and business appear to be completely incomparable. War is in essence destructive. It damages, breaks, kills and exterminates. By contrast, business is ideally constructive. It gathers, grows, produces and enlarges. Both are existential experiences to man, but they are not interchangeable, not even as complements. Instead, they exist at the same time, next to each other, and - as Tolstoy and Mitchell show - they can be juxtaposed in the most contradictory ways.

This is not only true for literary fiction, but it is also a phenomenon we can identify in many epochs of history. The Roman Empire, for example, came into being through a series of brutal wars, violence, slavery and the annihilation of rivals like the Carthaginian Empire. Nevertheless, the economy everywhere around the Mediterranean flourished and expanded to an extent which would have been impossible without the Roman campaigns. Conversely, unbalanced economic growth brought civil war and violent unrest to many countries during the industrial revolution with uncountable casualties. Such experiences of gain and loss, construction and destruction cannot be directly compared. They are distinct perspectives on human life.

4.   Explanation for the Coexistence of Frameworks
That art is based on a framework of thinking wholly different from other approaches to reality has been widely discussed since the modern has become a term of philosophical thought. The discussion was inevitable when the technical possibilities of image processing outgrew the capabilities of the painter. Photography was now able to reproduce the visual appearance of any motif much better than a painting made by human hand. However, it turned out that the technical perfection of photography could not insure that it depicted the essence of the motif. It became apparent that the role of art was more than reproduction, that art is able to express something logically indescribable. The modern, as Lyotard characterizes it, is the epoch of awareness of the indescribable in art, while the post-modern expresses the fact of indescribability itself.[3] Although this theory originated in art, it spread to practically all aspects of human culture in the twentieth century, including even physics and mathematics, which both found that there were limitations in the technological access to quanta and to logical expressions. It seems generally true that there are two frameworks of approach to the world, which can be called the technological framework and the essential framework. The technological framework covers the rationally oriented description. It is called "techno"-

"logical" because the description is achieved by some kind of machinery, whether this means a physical tool or the apparatus of logic itself. The essential framework includes the indescribable that is beyond the reach of technology. Approaches to the world within the essential framework are not justified by expressive logic but by intuitive acceptance.

From the point of view of modern thinking, it is therefore no surprise that we can identify two levels of approach to war and business. The surprising aspect is that the relationship between war and business turns out to be completely different depending on the framework in which they are viewed. It seems hard to believe that war and business can be treated almost the same within the technological framework, while the two phenomena are absolutely incomparable within the essential framework.

At this point there are two questions we should ask: First, how can there be such a difference in the treatment of war and business, depending on the framework we are in? And second, why can we get along with war and business fairly well without concerning ourselves with the framework? The key to the answer is the distinction between activity and experience. Both play an important role in human life, but once again in quite different ways.

Experience is a holistic process of confronting, managing and being influenced by life situations. It is obviously embedded in the essential framework. But experience is not the whole of living. Living requires activity, and activity is not necessarily referring to the essential framework. Most of our activities are to some extent controlled or overseen by rational thinking. Only in emergency situations or in moments of great emotional power is it possible for our instincts to take over completely without us having some idea of what is happening in the rational parts of our brain. Rational control and surveillance have many advantages. First, they let us reflect on and evaluate what we are doing. By this we become able to improve our behaviour, and also to justify and explain it to ourselves as well as to others. The aspect of justification makes clear that rational thinking is a fundamental requirement of socially coherent behaviour. Rational thinking enables us to understand the behaviour of others and for them to understand our behaviour, even if they are in a completely different situation than we are. The famous cognitive scientist Piaget showed how this ability, which he characterizes as "operational cognition," emerges in the individual development of each human, reaching its full extent not before the age of twelve, possibly even later. Rational thinking is not only a basis for social behaviour and learning; but it is also a very useful shortcut to reduce the amount of thinking we have to do before we act. If we had to meditate about all aspects of our current situation and wait until we get an instinctive conviction about what is best, our actions might be more natural, but most of them would certainly also be much too slow to make sense. The same idea of abbreviation is of course the background for the development of technical machinery, which emphasizes the suitability of the term "technological framework" introduced above.

5.   The Metaphor of the Game

Although activity always coincides with experience, it is not necessarily embedded in the essential framework of thinking. In fact, there is one activity which we explicitly expect is not connected to the essential framework: we call it a game. Games are considered to be happening without any consequences to "real life." Nevertheless, many games have an existential meaning, like the games of young animals which are at the same time exercise and an important preparation for their future lives. All games use up a part of our lifetime and they also shape a part of our experience. Calling a game an artificial activity seems, therefore, inadequate. However, many games take place in an environment which has not naturally emerged, but was created for no other use than playing the game. This is especially true for sports or social games as we see them in children's play, incorporating all kinds of common adult situations, such as raising a family, going to the doctor, being a soldier or doing business. To such games, it seems suitable to apply the term "virtual." Virtual games are leisure activities. When we play them, it is unnecessary for us to think within the essential framework. We can concentrate completely on the technological framework established by the rules of the game and the roles of the players.

It appears, moreover, that this is not only true for games in the narrower sense, but for a whole variety of human activities. Sartre suspected that all actions in life are more or less like playing a game. He said that perhaps we must choose, either to be nothing, or to play what we are.[4] Wittgenstein also used the notion of games in the context of language. He noticed that most of the things we say are not based on a full logical concept of reality. The logical concept behind language is more likely some kind of game with a limited set of abstract rules.[5]

Therefore the answer to the second question raised above is that we apparently perform war and business, like many other activities, as games. In war and business we usually act in a virtual environment which does not need to refer to those pursuits' essential meaning for human life. Consequently, the scenario of a game can establish general structures for both activities without any problem.

Concerning the first question mentioned above, there is good reason to assume that modern society benefits quite substantially from the possibility of subsuming war and business under similar kinds of structures. Therefore living with such contradictions makes life actually easier than trying to keep our thinking consistent.

6.   Historical Development towards Inconsistency

It was mentioned earlier that the distinction between technology and essence became inevitable in art when the quality of technical reproduction rose to continuously higher grades of perfection. Similarly, the importance of technological thinking in war and business does not seem to be innate. It is more likely the result of the historical development of the last four hundred years. During this period of time, both war and business became much more measurable and mechanistic. Warfare had to change because the introduction of firearms required better coordinated procedures for making war. In former times the soldiers, for the most part, simply ran forward and attacked the enemy. The strategic impact of the general was limited to the decisions of where and when to strike and of which part of the army should proceed and which should fall back.[6] In the seventeenth century, a new style of fighting emerged, with soldiers in clear rows approaching each other from a distance, one row shooting while the other one was reloading. Uncoordinated man-to-man combat only started at the end of the advance - if it was still necessary after the shooting. This kind of warfare was much more effective, but it required extensive training and drill to make sure that the soldiers did not shoot their comrades instead of the enemy.

At roughly the same time in history, the financial system became more elaborate, changing the way that business was done. Before that period, payment of goods was based on a rough estimate of their value, mostly in gold or other precious material. Now, stock exchanges and banks became popular. They made it possible to measure the exact value of products and services in the market, the benefits of different payment methods and of interests, and the impact of taxes.

As a result, extensive calculation and planning became an increasingly crucial prerequisite for success for both business and war. The mathematical background for both was the game theory introduced by Pascal and Fermat around 1650. Game theory made it possible to evaluate the quality of decisions on the basis of probabilities. This theory was expanded in the nineteenth century to all kinds of scenarios, from ballistics to agriculture. However, it too remained relatively undeveloped until the first half of the twentieth century, when work on these topics was gathered and systematized by a general approach called Operations Research.

7.   Operations Research

Operations Research can be described as the study of the usage of mathematical models, statistics and algorithms to aid in decision-making. It analyses complex real-world systems, typically with the goal of improving or optimising performance. The term "Operations Research" is often used in close conjunction with "Management Science." However, Operations Research does not deal solely with business; it also applies to any other rationally describable behaviour, especially warfare. The development of Operations Research was closely related to the Cold War. Militaries in the East and the West were extremely interested in devising a scientific basis for their operations, which had acquired greater complexity with the ongoing rise of technical warfare. This line of development in the West was best represented by the career of Robert McNamara, who left the presidency of Ford Motor Company and, as Lyndon Johnson's Secretary of Defence, applied scientific management theories to the prosecution of America's war in Vietnam. War and business can therefore be seen as the two most prominent and important fields of application for Operations Research. Other scenarios like sports or social science play only marginal roles.

In business and war, Operations Research took over the role that photography had taken for painting, making it possible for scientific methods to calculate the decisions much better than any human instinct could. Operations Research can calculate which nodes in the enemy's military system are the most vulnerable and how to strike them to cause the most devastating and long-lasting effect. It can also calculate which distribution process for a certain product is fastest and most effective and reaches the most customers. The methods used are identical in both cases. In fact, results were often transferred from one scenario to the other. For example, a theory developed for the improvement of the Russian railway system in the 1930s, which applied several new mathematic theorems, was almost exactly copied more than twenty years later for the strategic planning of the most effective bombing of the Soviet Union.[7] Conversely, the design of the Internet, which was originally a secret military network, became, as we all know, the basis for a whole new kind of business activity.

However, the application of Operations Research to warfare and business lacked any reflection on the essence as is practiced in the arts. The reduction of war and business to the virtual environment of the technological framework is mostly considered a simple and consequential process. The missing respect for the essential framework has not been expressed openly. Instead it has led to a general feeling of uneasiness, which tends to brand war - and in a rather similar way business as well - as inhuman. Yet what else, it might be asked, could war be than a completely human issue?

8.   Improving the Efficiency of Activities

The ability to play a game is connected with a relatively early stage of cognitive development which Piaget attributes to children at the age of four. At this age, the human mind learns how to adopt a formal set of rules and behave according to them. Every soldier or businessperson is therefore basically able to assume a role. Obviously, assuming the role of the soldier and persevering in it requires much less effort than thinking again and again about potentially risking one's own life and killing one's fellow man. If a soldier continuously considered all the existential implications of his actions, he would probably never be able to stand his ground. A businessperson might not operate in such extreme situations as a soldier; but the principle is the same for a manager who becomes burdened with too much work and suffers a heart attack as well as for a factory worker who spends day after day making the same monotonous movements. Continuous consideration of the value of their work to their personal existence would hardly make them behave as they do. The challenge in both war and business is thus not so much making people assume roles, but rather making clear to them what roles they are supposed to assume. Propaganda, hierarchies, human resource management and advertising are all mechanisms to tell people what their roles are. None of them changes people into marionettes. There is no interest in controlling behaviour in full, but rather in establishing the rules of the game and assigning the roles that are to be played. Field soldiers, spies, generals, merchants, factory workers and managers are some of the most clearly defined stereotypes in our culture. They are common supporting characters in cinema and television. Any actor is able to impersonate each one of them by showing some typical behaviour. In fact, today's warfare concepts and business models seem to be trying to fit into the roles designed for them more than ever before. Advertising does not stop by telling us how to be a customer, but also by showing us how accountants, merchants and workers behave. Terrorism as the most extreme form of propaganda tries to force people into armed conflict by leaving them no other choice than fighting for one side or the other. Both are contradictory to the freedom of democracy, where everyone is allowed to decide as much as possible on their own which game they want to play.

9.   Neither Framework Can Exist without the Other

Although thinking can flow freely between the technological framework and the essential framework, this does not imply that thinking as a whole can be reduced to only one of them. There is no doubt that culture and society necessarily involve advanced forms of communication, role-playing and conventions about values, and all this is impossible without the technological framework. A socially active and cultured person cannot live without the technological framework. It therefore seems possible to reduce one's thinking to the technological framework, since, as has been noted, decisions can be more easily made with exclusive respect to this framework. However, although decision-making may be a prerequisite for rational action, it does not dictate the action as a whole. Like many other scientists of the twentieth century, Wittgenstein gained this insight after a long and painful research process. His understanding was strongly connected with his notion of the game. Wittgenstein introduced the game in his theory because he had reached the conclusion that explicit symbol logic was not sufficient to explain the practical use of human languages. He understood that speaking a language is like playing a game of chess. In contrast to card games, for instance, chess has no hidden information. Both players see the setting and they know the rules. However, while the setting and the rules can explain decisions, they are not sufficient to explain the activity as a whole. They are the rationalized part of the game, but not the substance. With this argument, Wittgenstein touched a point which was to become a major issue in information technology. Even today, chess computers are still not necessarily superior to human brains, and in robotics researchers have learned that it is rather difficult to make machines play soccer: they are just not interested enough in getting the ball into the other team's net.[8] Operations Research is, of course, very interested in correcting such flaws as much as possible. This can be achieved by establishing new layers of technology on top of the others, like meta-games which refer to another game and not to the abstraction of a real situation. Finance is a good example. At first, the game of exchanging money replaced the exchange of real goods. The exchange of money was then again covered by other layers, exchanging cheques instead of coins, then options, interest rates, expectations of market growth and more. But even if machines play a more and more important role on each additional level and the completeness of the rules of the game is growing as well, finance as a human activity is far removed from rational behaviour.[9] In the same way, new weapons and strategies have not made war in any way more rational. Quite the contrary, combat has become a most successful simulation scenario for computer games because it offers so many degrees of freedom to make the game more interesting.

10. Conclusion

  My intention in this paper has been to show that there are two different frameworks of thinking within which we can approach war and business. I have shown that the relationship between war and business is quite different depending on the framework to which we are referring. War and business can be treated very much in the same way within the technological framework of thinking, but they are practically incomparable within the essential framework. Both frameworks are deeply woven into human life. While the essential framework is connected to holistic experience, the technological framework is connected to activity. The best example of the fact that activities need not possess an existential basis is the game, which applies as a metaphor to almost all kinds of cultural and social behaviour. Such activities can be called virtual. Certainly war and business can be included among them. Acting on a virtual basis can be much more effective and easy than carrying the load of existential questions and meaning at all times. Thus it is no surprise that there is a growing tendency to reduce our thinking to the technological framework in modern society.

  However, it is dangerous to underestimate the role of the essential framework in our thinking merely because it appears to be not necessarily involved in rational decision-making and acting. Even if we are not consciously aware of the influence of the essential framework on our thinking, it can sooner or later have a significant impact on our lives. Freud's psychoanalysis, which appeared in parallel with modern art and shared much of the idea of the indescribable with that movement, was the first to identify unconscious thoughts as an important aspect of our personality. The unconsciousness is able to cause severe psychological damage if too much existential experience is repressed. It is hardly a surprise that war experience and business pressure alike are major causes of neuroses and depression, and both have become common diseases in modern society. The repression of thoughts is apparently a greater danger to health than the knowledge of inconsistencies in one's rational thinking. Echoing Nietzsche, Paul Valéry stated that conflicting ideas and convictions are not uncommon in modern society.[10] It seems therefore quite natural to live with the fact that war and business can have a completely different relationship to each other, depending on the framework of thinking. Both scenarios of thinking - the technological, where business and war can be treated in the same way, and the essential, where they are completely incomparable experiences - have good reasons to exist. There is nothing wrong with keeping them both in mind at the same time.

Notes

[1] Sun Tzu's original text is relatively short and therefore usually amended with lots of additional comments aimed at the special interest of the readers. See Gerald A. Michaelson, *Sun Tzu: The Art of War for Managers* (Avon, MA: Adams Media Corporation, 2001) for its extensive comments and interpretations. An important complementary text to Sun Tzu is Carl von Clausewitz, *Vom Kriege* (Reinbek: Rowohlt, 1963). See also Kenneth C. Allard, *Business as War: Battling for Competitive Advantage* (Hoboken, NJ: J. Wiley, 2004).

[2] Leo Tolstoy, *Woina i Mir* (Moskow: Literaturnoje Nasledstwo, 1983); Margaret Mitchell, *Gone with the Wind* (New York: Scribner, 1996). For an interpretation of Tolstoy's portrayal of the complexities of human nature see Isaiah Berlin, *The Hedgehog and the Fox: An Essay on Tolstoy's View of History* (New York: Simon and Schuster, 1953).

[3] On the question "What is postmodern?" see Jean-François Lyotard, "Beantwortung der Frage: Was ist postmodern?" in *Postmoderne und Dekonstruktion. Texte französischer Philosophen der Gegenwart*, ed. Peter Engelmann (Stuttgart: Reclam, 1990).

[4] Jean-Paul Sartre, *Die Wege der Freiheit* (Reinbek: Rowohlt, 1949).

[5] Ludwig Wittgenstein, *Philosophische Untersuchungen* (Frankfurt am Main: 1984).

[6] An extensive description of the fighting traditions in different cultures and epochs can be found in John Keegan, *A History of Warfare* (New York: Knopf, 1993).

[7] See A.N. Tolstoy, "Metody nakhozhdeniya naimen'shego summovogo kilometrazha pri planirovanii perevozok v prostranstve," in *Planirovanie Perevozok* [Transportation Planning] 1 (1930): 23-55. The title in translation is: "Methods of finding the minimal total kilometrage in cargo transportation planning in space." See also L.R. Ford and D.R. Fulkerson, *A Primal Dual Algorithm for the Capacitated Hitchcock Problem (Notes on Linear Programming: Part XXXIV, Research Memorandum RM-1798 (ASTIA Document Number AD 112372))* (Santa Monica: The RAND Corporation, 1956). There is reason to assume that Ford and Fulkerson had heard of Tolstoy's paper. However, Ford's other works cover so much basic theory on the max flow/min cut approach to optimisation that he can be trusted to have developed much of the theory on his own, even if it is similar to what Tolstoy wrote.

[8] This became apparent during the first RoboCup Championships for robots playing soccer, which have been held regularly since 1996. Sometimes the teams do not move at all; see for example Gunter Görz and Bernhard

Nebel, "Künstliche Intelligenz" [Artificial Intelligence]. Unpublished paper presented in Frankfurt am Main, 2003.
[9] See Dietrich Dörner, *Die Logik des Misslingens: Strategisches Denken in Komplexen Situationen* (Reinbek: Rowohlt, 1989); and H.A. Simon, "Rational Choice and the Structure of Environments," *Psychological Review* 63 (1956): 129-138.
[10] Paul Valéry, "Die Krise des Geistes," in *Werke*, ed. Jürgen Schmidt-Radefeldt (Frankfurt am Main: Insel, 1995).

Inventing the General:
A Re-appraisal of the *Sunzi bingfa*

Andrew Meyer and Andrew R. Wilson

Abstract: The *Sunzi bingfa*, commonly known as *Sun Tzu's Art of War*, reads like a treasure trove of strategic wisdom. When it first appeared in the late fourth century BCE, however, the *Sunzi* drew intense censure for its radical departure from the strategic culture of the age and for the author's call for a professional standard of command in war. The revolutionary nature of the text is often lost on contemporary readers who assume the existence of a professionally-run military. This oversight obscures much of the potential value of this book for modern students of war and strategic theory and leaves the text open to severe misinterpretation. The mass armies common in late fourth-century BCE China could not be effectively led until new social roles were created, such as military officers who wielded routinised rather than charismatic authority. The *Sunzi* railed against "traditional" approaches to war and argued for a strategic culture centred upon the professional expertise of the commander. In other words, the author was inventing the "general" and providing the conceptual framework within which the military technology of his day could reach its full potential. The *Sunzian* general, for whom command was not a test of valour or mantic office but an intellectual enterprise, was defined by his professional expertise and unique qualities of mind. As such, the *Sunzi* is an elaborate defence of the authority of the commander and the autonomy of the realm within which he operated. The *Sunzi*, therefore, anticipates the kind of military-intellectual complex that all advanced societies manifest, and highlights many of the enduring tensions evident between the "professional" military and "amateur" statesmen.

Key Words: Sun Tzu, Art of War, Command, Confucius, Strategic Theory, Ancient China, Warring States

1.   Introduction

Now the ways by which the ruler brings calamity to the army are threefold: Not understanding that the army may not retreat yet telling it to retreat, this is called misleading the army; not understanding the affairs of the three armies and taking over their administration, this will confuse the officers of the three armies; not understanding the powers of the three armies and assuming their responsibilities, this will cause the officers of the three armies to doubt. When the three armies are in doubt and confused, the impositions of the feudal lords will arrive. This is called disordering the army and inviting conquest.

Sunzi[1]

There are orders of the ruler that are not obeyed.

Sunzi[2]

A general whose genius and hands are tied by an Aulic Council five hundred miles distant cannot be a match for one who has liberty of action, other things being equal.... [I]nterfered with and opposed in all his enterprises [he] will be unable to achieve success, even if he have the requisite ability. It may be said that a sovereign might accompany the army and not interfere with his general, but, on the contrary, aid him with all the weight of his influence.

Jomini (1779-1869)[3]

Politics uses war as a means to achieve its own ends. It exercises a decisive influence on the beginning and on the end of war... It is best, therefore, for strategy to work closely with politics and only for political goals, but in the action it prescribes to be totally independent of politics.

Helmuth von Moltke, the Elder (1800-91)[4]

Politics and strategy are radically and fundamentally things apart. Strategy begins where politics end. All that soldiers ask is that once the policy is settled, strategy and command shall be regarded as being in a sphere apart from politics... The line of demarcation must be drawn between politics and strategy.

U.S. Army Command and General Staff School (1936)[5]

The *Sunzi bingfa* (*Sun Tzu's Art of War* as the title is usually rendered) is perhaps the most frequently cited work of strategic theory in the world. Despite its composition over two thousand years ago, during

China's Warring States Era (453-221 BCE), it remains required reading at most military academies and staff colleges in the western world. In addition, new translations of this terse classic appear every year in dozens of languages and there is a cottage industry of "Sun Tzu experts" who apply the Master's adages to the full-range of the human experience, from romance to finance to sports.

While our purpose is not to question the obvious allure of the *Sunzi*, we nonetheless seek to encourage students of the text to balance their enthusiasm for its relevance to the contemporary world with a measure of appreciation for what the author was trying to accomplish in his day. The obvious sagacity of the truisms that fill the *Sunzi* makes it a useful tool for would-be strategic theorists who want to validate their own theories with reference to the classics. While there is nothing inherently wrong in reading the *Sunzi* from these sorts of perspectives, taking the text as a kind of cookbook for victory without regard to the rhetorical and polemical purpose for which it was initially written leaves it open to promiscuous misapplication.

2.   "Master" Sun: History as Rhetoric

The *Sunzi* is popular for its brevity, its directness, and the eternal applicability of the platitudes it contains. The standard text comprises thirteen essays attributed to a "Master," a *zi*. Given that they are the words of a Master, they appear to be statements of truth, free from such extensive supporting analysis as one would find in Clausewitz's *On War* or other Western strategic theories. In addition, the text is crafted in the terse and stylised literary language of the late Spring and Autumn Period, circa 500 BCE. As such, the truisms read (in both Chinese and English) as discrete pearls of wisdom that are easily divorced from the rest of the text. In actuality the *Sunzi* is a coherent, logically organized and intellectually challenging strategic theory. The arguments build to a rhetorical crescendo, the language is consistent and purposeful, and the author mercilessly pushes an ideological and professional agenda. Contemporary readers, however, rarely appreciate this instrumental design of the text.

The author of the *Sunzi bingfa* makes three interrelated arguments. The first is the obvious claim - embodied in the title and in the opening line of each chapter, "Master Sun said" - that the text contains the strategic and operational wisdom of Sun Wu, a general who led the armies of the state of Wu to victory in the late sixth century BCE and who deserves the honorific of Master (*zi*). The second claim is that the sole purpose for the existence and employment of the military (*bing*) is primarily to defend, but also ultimately to increase the wealth and power of the state. And third, is that the professional general can and must wield the military with the same skill (*fa*) and degree of autonomy that a master swordsman handles his own weapon. In fact there is a running metaphor throughout the thirteen chapters emphasizing the intimate connection between the *jiang* (general or literally "wielder") and the *bing* (military or literally "edged weapon"). We can see these three claims interwoven in stark relief in the first line of the text.

> Master Sun said: The use of the military is the greatest
> affair of the state. It is the terrain of life and death, the
> path of survival and ruin, it must be studied.[6]

From a twenty-first century perspective this assertion seems axiomatic, but when it was written this was a highly charged and revolutionary statement.

Both the famous general to whom the book is attributed and the text itself were products of different historical moments within what is generally termed the Eastern Zhou. The Zhou era, named after the dynasty that came to power in the eleventh century BCE, is divided into two main periods: Western Zhou, 1045-770, and Eastern Zhou, 770-256. Following the sack of the Zhou capital of Han, near modern-day Xian, in 770 the dynasty relocated to the Eastern city of Luoyang. During the Western period the Zhou can be characterized as the dominant clan and the dominant state in a feudal coalition. The loss of the West, however, deprived the Zhou of the fundaments of real power. Therefore the Eastern Zhou period was characterized by a contest among ruling households to determine which would serve as protector (hegemon king) of the enfeebled dynasty. This contest escalated in the Spring and Autumn Period (722-481) and peaked in the Warring States Era (403-221).

In the Spring and Autumn, the various dukedoms and principalities that made up the Eastern Zhou political order were decreasing in number but increasing in size. In this same period war was beginning the transition from seasonal, stylised and ritualistic combat among chariot-mounted aristocrats to an age of mass infantry armies, spurred by demographic surges, bureaucratic and technological innovations, and the introduction of cheap horsepower from the steppe.

This is the moment when General Sun Wu emerges. Sun Wu was born in the northeastern state of Qi, which had a proud martial tradition; but it was in the service of the emerging southern state of Wu that he came to prominence.[7] Wu had long been harassed by its more powerful western neighbour, Chu, but through better strategy and preparations had succeeded in blunting Chu's attacks and was planning to go on the offensive.[8] King Helü of Wu had the services of an able minister, Wu Zixu, himself an exile from Chu, whose greatest gift was his eye for talent.

Around 509 BCE, Wu Zixu convinced the King to entrust the armies of
Wu to Sun Wu, after which they achieved a dramatic victory. That victory
placed the King of Wu among the ranks of hegemons, i.e. protectors of the
largely powerless Zhou kings. After this point, however, Sun Wu's brief
role in the tragedy of Wu ends and the dramatic focus returns to Wu Zixu
and the King of Wu.

Riding the crest of its victory over Chu, Wu overawed its
neighbours. But when King Helü died in 496 his successor, Fuchai,
increasingly ignored Wu Zixu's advice (and ultimately ordered his
execution). Thus without expertise and wise guidance, the state of Wu was
ground down in a series of protracted wars, in the operational planning of
which Fuchai himself meddled. This decline allowed the larger state of
Yue to build up sufficient power to finally destroy Wu, its long-time
enemy, in 473; but in its turn Yue, having been weakened by these
protracted conflicts, was ultimately destroyed by a resurgent Chu.[9] To
anyone reading the *Sunzi* when it appeared in the late fourth century, this
great tragic drama of Wu, and the personalities involved, would have
come immediately to mind.

**3.   "The Supreme Excellence": Overcoming the Aristocratic
         Ethos of the Bronze Age**
In the two centuries that followed the demise of Wu, war
accelerated in scale, breadth and lethality, and the pretensions of an earlier
age were rapidly unravelling. The Warring States Era saw ever larger and
more organized states begin to consolidate, so that by the beginning of the
fourth century, only eight or nine very large states remained, all vying in a
zero-sum contest. This was when the *Sunzi bingfa* appeared (in other
words was "found"), a context that saw the rise of states large and lethal
enough to vie for absolute mastery over all of ancient China.

One would think that such dramatic changes in the scale and
scope of warfare, with armies of several hundred thousand men armed
with standardized weapons and requiring elaborate logistical support,
would have necessitated a dramatic re-appraisal of strategic culture and of
the military profession.[10] In fact, the author of the *Sunzi bingfa* was
contending with a military culture still emerging from the aristocratic era.
Though new technologies and new forms of organisation were being
deployed on the field of battle, the same warrior aristocracy of the earlier
Bronze Age still monopolized the reins of power, and old aristocratic
values still prevailed. The aristocratic warrior operated under imperatives
completely different from the military commander the *Sunzi* is attempting
to invent. For the aristocrat two qualities were paramount: gentility of
birth and valorous conduct. Without these two qualities the aristocrat lost

all legitimate claims to power, and thus the aristocracy would defend them
at all cost. An old adage aptly describes how "sacrifice and warfare" were
the two great affairs of the aristocratic warrior. Sacrificing to his ancestors,
the aristocrat publicly proclaimed his hereditary right of leadership and
celebrated the ties of kinship that bound him to his ruler and (more often
than not during the Bronze Age) to the warriors under his command.
Warfare provided the aristocrat with an arena in which to demonstrate his
personal valour. By risking his life in open combat he displayed the virtue
he had inherited from his ancestors. By shedding his enemy's blood he
performed a further act of sacrifice that redounded to the glory of his
ancestors in the afterlife. From the perspective of the aristocratic warrior,
the *Sunzi*'s injunction that

> To achieve one hundred victories in one hundred battles
> is not the supreme excellence, to reject battle and yet
> force the submission of the enemy's troops is the
> supreme excellence. [11]

is worse than meaningless, it is offensive. Victory without combat
precluded a display of martial virtue and completely subverted the very
point of taking up arms in the first place. It is this strategic and martial
culture that the *Sunzi bingfa* critiques.

4.   Sun Wu and Confucius: The Resistance of the "Masters"

In this antipathy between the perspective of the *Sunzi* and that of
the aristocracy, one can see one of two pressures driving the text's author
to couch his critique in artifice (claiming Sun Wu as the voice of wisdom)
and purposive anachronism. The other source of pressure came from the
group he purported to join, the authors of the growing body of Masters'
texts. The Masters were writing in response to the collapse of the
institutions of the Western Zhou and the Spring and Autumn, and like the
author of the *Sunzi* they strove to demonstrate how the social and political
institutions of the Sinic world could be restored to harmony (or at least
rationality). The Masters were also disposed to be critical of the
aristocratic ethos of the Bronze Age.

Few of the Masters were men of unimpeachably high birth; they
donned the cloak of a new kind of authority based on wisdom and insight
rather than heredity and martial valour. The term they appropriated to
themselves, "Master" (*zi*) underscores the sense in which the Masters were
engaged in a process of social invention identical to that of the *Sunzi*'s
author. Originally "*zi*" was the second-to-lowest title in the hierarchy of
hereditary ranks of the Zhou. [12]   The consolidation of sovereign regimes

during the Warring States had made the lower orders of the aristocracy largely moribund for practical political purposes. The Masters thus appropriated what had become a courtesy title among low-level aristocrats and redefined it to mean something like "Doctor" or "Professor." In this way the Masters asserted that they could stand shoulder to shoulder with the aristocracy while still insisting that their authority had its own autonomous source and logic. The first Master was Confucius, or Kong Fuzi, and his attitude towards military affairs was typical of the Masters who rose up after and in imitation of him.

> Duke Ling of Wei asked Confucius about military formations. Confucius answered, "I have, indeed, heard something about the use of sacrificial vessels, but I have never studied the matter of commanding troops." The next day he departed.[13]

Despite the centrality of interstate conflict in Ancient China, it was not entirely clear that the raising and employment of armies was a fit topic for masterly discourse. Because men like Confucius lacked the pedigree of the rulers of the Warring States they were always on the defensive to justify their authority as planners for state and society. The literary medium in which they formulated and expressed their thoughts was considered the antithesis of the martial realm; thus it became a standard theme for writers like Confucius (insulted as he is by the Duke's question) to denigrate martial pursuits as either inferior to the arts of peace or wholly aberrant.

5.   *Junzi* and *Jiang*: The Struggle to Create New Forms of Authority

One can see both why the author of the *Sunzi* would want to latch on to the new position of authority carved out by the Masters and why his emergence onto the stage of the Masters' discourse would be most unwelcome by the Masters themselves. Just like the other Masters, the author of the *Sunzi* was attempting to carve out new social roles and forms of authority to meet the needs of the new age. This is well illustrated by a comparison of the treatment of the term *junzi* by Confucius with that of the term *jiang* by the *Sunzi*: terms that had potent conventional valences before their appropriation by these two respective Masters.

Junzi was the closest equivalent in ancient Chinese for our English word "aristocrat." Its two parts broke down into "ruler" (*jun*) and "son" (*zi*), implying that the minimum criterion for aristocratic status was descent from a ruler somewhere back in one's family tree. The number of

genuine "rulers" was far less during the Warring States Era than it had
been in previous centuries, but the reigning assumption remained that all
genuine "aristocrats" could trace their ancestry back to some *bona fide*
ruler of the Bronze Age, even if one's relationship to the ruling lineage
was indirect or the state ruled by that lineage no longer existed.

Confucius was one such claimant to aristocratic status; one would
have to go far back in his family tree to find a genuine ruler, but could
eventually find some of the ancient rulers of the state of Sung. Despite his
legitimate claim to minimal aristocratic status, Confucius was not inclined
to accept the logic of the aristocratic ethos. He claimed and expected to
exercise greater authority than his pedigree of birth would ever warrant.
For this reason (at least in part) Confucius set out to completely redefine
the valence of the term *junzi*. In his teachings, *junzi* is drained of virtually
all the content it conventionally held and was imbued with new
significance specific to Confucius and his ideals. Take for example, this
passage in the *Analects*:

> In Chen, when provisions had run out the followers had
> become so weak that none of them could rise to their
> feet. Zilu, with resentment written all over his face, said,
> "Are there times when even gentlemen [*junzi*] are
> brought to such extreme straits?" The Master said, "It
> comes as no surprise to the gentleman [*junzi*] to find
> himself in extreme straits. The small man, finding
> himself in extreme straits would throw over all
> restraints."[14]

Conventional logic is entirely with Confucius' lampooned
student, Zilu. Well-fed leisure was a hallmark of aristocratic status; thus
for a group of men so low on the aristocratic ladder to find themselves
starving would seem to indicate that they had slipped over the edge into
commoner status. Confucius' insistence that they may still call themselves
junzi necessitates a complete redefinition of that term. In Confucius'
teachings *junzi* becomes a moral rather than an inherited status. A *junzi* is
not defined by the objective conditions of his birth or his economic
circumstance, but rather is to be identified by his subjective response to
those conditions. A *junzi* is not someone who remains well-fed despite the
fact that he does not work; rather he is a person who remains composed
and unafraid despite the fact that he is not fed.

We can see that the author of the *Sunzi* is redefining the word
jiang in the same manner that Confucius set out to redefine the term *junzi*.

A typical pre-*Sunzian* use of the word *jiang* can be found in the *Zuo zhuan*, a chronicle of the Spring and Autumn Period:

> The King had the Central Army. Duke Linfu of Guo led the Right Army [*jiang*], which included the men of Cai and Wei. Duke Heijian of Zhou led the left army [*jiang*], which included the men of Chen.[15]

In the Western Zhou and in the Spring and Autumn, "command" of an army was a wholly interchangeable task passed from aristocrat to aristocrat as the occasion or the whims of the monarch demanded. Here we can see that overall command goes to the Zhou king himself, and the choice of whom to command the armies on his flanks is dictated by the highborn status of the two men and their closeness in kinship to the king himself. The difference of this type of command and the armies over which it was exercised from that of the Warring States is obvious. The three "armies" to which this quote refers are really just three groups of chariots arrayed at different positions on the field.[16] Taking "command" is thus merely a function of "grasping" [*jiang*] the reins of the lead chariot and requires just as much skill. Compare that notion of "command" to the use of the term in the *Sunzi*:

> Thus the commander [*jiang*] who understands the military is the arbiter of the fate of the people, the master of security and peril of the kingdom and the dynasty.[17]

The *Sunzi*'s concept of *jiang* is starkly different from the conventional Bronze Age usage. In the text the commander has risen from a position to be held by a man of good breeding to one of godlike powers that must be filled by an individual possessed of extraordinary knowledge and skills.

Given this congruence of goal and purpose in the teachings of the *Sunzi* and Confucius, we can sense the potential for controversy. While both teachers were attempting to realign values and terms, the *Sunzi* was attempting to give new priority and autonomy to a realm that Confucius would see denigrated in his project of social reinvention. Knowing the importance of the martial realm to the aristocracy's claims of superiority, we can well understand the resistance of someone like Confucius to the *Sunzi*'s argument that the professional management and employment of mass infantry armies is not just a fit topic for Masters, it is the greatest affair of the state!

This intrinsic dynamic makes Sun Wu an exceptional choice for author of such a text. Not only is he a convenient cipher for the later author of the *Sunzi*, conveniently mute and only vaguely portrayed in the existing sources (the exact details of his generalship are not discussed), but most importantly he possesses the invaluable asset of being Confucius' contemporary. In a culture where legitimacy was quantifiable in terms of antiquity, this lends credibility to his assertions and to the author's bold claim that Sun Wu deserves the honorific of "Master." Moreover the meteoric rise and spectacular collapse of Wu would have hung over everyone reading this book, thus creating instant recognition and resonance for the readers. But whereas Wu Zixu, the prime minister, was traditionally seen as the more dynamic and important agent of Wu's rise, giving Sun Wu a voice asserts that military leaders are neither passive nor voiceless and that military leadership possesses its own autonomous realm of knowledge and concerns. By attributing the book to Sun Wu the author is thus at liberty, from the legitimacy of antiquity, to launch a polemical assault on the strategic culture of the Warring States and the endurance of anachronistic attitudes.

6.   Inventing the General: A New Ideal for a New Age
The mass armies equipped with standardized weapons that were common in late fourth century China could not be effectively led until new social roles were created, such as military officers who wielded routinised rather than charismatic authority. The *Sunzi* thus railed against "traditional" approaches to war that dominated Warring States' military thought, and argued for a strategic culture centred instead upon the professional expertise of the commander. The polemic is so rhetorically extreme, especially in its demands on the general for a coldly rational and anti-heroic approach to war, based on a nearly perfect knowledge of the adversary and near-total command of the physical environment, because the attitudes the text is attacking were so deeply entrenched:

> If a general who heeds my assessments is used it is
> certain victory, retain him! If a general who does not
> heed my assessments is used it is certain defeat, remove
> him![18]

By inventing the superior "general" the author is also implicitly providing the conceptual framework within which the military technology of his day could reach its full lethal potential.

The *Sunzi* presents a revolutionary ideal to which the professional commander must aspire and against which he is to be measured. While the ideal may not be wholly realistic, its practical utility far surpassed the old heroic/aristocratic ideal given the new nature of warfare. Unfortunately, the vast majority of people who read the text today assume that its rhetorical extremes are literally presented as practical advice. Instead, the author is first establishing a new ideal of professional command as an intellectual enterprise and secondly is attempting to starkly delineate his realm of professional expertise. Two quotations should suffice to demonstrate these points:

> Therefore when victory is divined in the temple before battle, it is because one has drawn the majority of stalks. When defeat is divined in the temple before battle, it is because one has drawn fewer stalks. The majority is victorious, the few are defeated. How much more so with no stalks at all!?!?[19]

This passage refers to a ritual performed to divine the outcome of a battle in the royal ancestral temple involving the casting and drawing of yarrow stalks. Within the ethos of the aristocratic age, such a ceremony was indispensable because victory was assumed, in the final analysis, to be in the hands of spiritual forces. To fail to consult the spirits in the temple would have been an unforgivable act of hubris. Moreover, initiating the campaign with this type of divination stressed the notion that war was from beginning to end a ritual occasion; one went to war for the honour of one's ancestors and dedicated the fruits of the battlefield to their glory.

The author of the *Sunzi*'s point is two-fold. First, the use of armed force is a materially quantifiable process with a materially measurable outcome; therefore a careful examination of its fundamental elements is critical when mobilizing and employing mass armies. The glory of the ancestral spirits might be well and good, but if the logistical assets of the state fall below a certain critical number, the ancestral spirits will have no home or descendants of which to speak. Second, the *Sunzi* denies that war is a primarily spiritual or mystical endeavour and insists that success or failure resides wholly within the compass of human intelligence. He proves this by showing that even in the temple the ritual follows the same logic that inheres on the battlefield; greater numbers win. During the Bronze Age, the reigning assumption was that the course of the battle was shaped by the mysterious and inscrutable purposes of the spirit-world. The author of the *Sunzi* insists that this relationship should be inverted, and that the workings of the spirit world may be understood through studying the fundamental rules of the battlefield.

Ironically, given the starkly secular perspective manifest in the quote above, the *Sunzi* sets the new general apart from the traditional metrics of command and imbues him with almost supernatural qualities of mind:

> In general, these are the four ways of manoeuvring
> troops to seize the advantage and the means by which
> the Yellow Emperor vanquished his four rivals.[20]

What the author does not need to explain is that the Yellow Emperor was a god, as were his adversaries. Thus in this passage, and in numerous others, the new general is possessed of god-like abilities, further distancing him from mere "mortals" by asserting the uniqueness of the skill-sets involved and the extent of the specialized knowledge required to use the new military. The author is also clearly situating himself on one side of the debate over the relations between gods and mortals that raged in early China. Rather than trying to plumb the divine through omens, the *Sunzian* general becomes divine and imposes order on the infinite complexity of war.[21] The text even goes so far as to call on the general to proscribe divination in the ranks so that he and he alone is seen as the ultimate arbiter of victory and defeat.

The extremity of the *Sunzi's* exaltation of the general raises the suspicion that the text is caught in a paradox. Though it is coolly rational in discussing the influence of the spirit world on combat (or lack thereof), it launches into flights of fancy when describing the preternatural faculties of the commander. The contrast, of course, is rooted in the polemical aims of the text. The *Sunzi* sets out to establish a new ideal of command, one imbued with the same dignity and power of the old aristocratic hero. Though the godlike effectiveness of the *Sunzian* commander may not be pragmatically approachable, it is no less so than the hyperbolic valour of the aristocratic hero, and attempting to even approximate the godlike will yield vastly better practical results.

7.   The Genius of Command and Command as Genius

With this perspective in mind we can now read the *Sunzi* as both an elaborate assault on amateurish, vain aristocrats and on preachy Masters, and at the same time as a spirited delineation and defence of the realm of authority and expertise of the professional general. Command is no longer a test of valour or aristocratic virtue; it is an intellectual enterprise. This fundamental assertion explains the structure of the book: we are introduced in Chapter I to the critical calculations that must be made prior to war; Chapter II then deals with the direct costs and dangers of war; the subsequent chapters describe the nearly super-natural ability of the commander in understanding and organizing the troops, manoeuvre, knowledge of the terrain and weather, shaping his force, shaping the enemy; and finally the culminating Chapter XIII explains that all of this is possible through the use of spies.

The intense novelty of the *Sunzi*'s new concept of command can be perceived in its relentless attention to the general's role in gathering and analysing intelligence. This function so much defines the commander that, as the *Sunzi* ingeniously perceives, one of the most valuable sources of intelligence is concentrated in the person of the commander himself. Every move the commander makes potentially reveals insight into his plans, making it imperative that he physically remain out of sight:

> For this reason on the day the order is given, close the passes and break the tallies, do not admit his (i.e. the enemy's) emissaries. Stay strictly to the upper halls of the temple in prosecuting the affair. If the enemy opens the door you must rush to enter. Anticipate what he cherishes; misdirect him as to the time, pace the marking line and follow the enemy in deciding the affair of combat. For this reason, be like a virgin girl at the beginning. When the enemy opens a door, be like a fleeing rabbit, the enemy will not be able to withstand you.[22]

There is some humour to the injunction to "be like a virgin girl at the beginning," but the instruction is meant quite literally. Like a virgin girl hidden from public gaze until the day of her marriage and in radical contrast to the aristocratic warrior of old, the new commander should remain indoors. Prior to the army's deployment his task is to process intelligence whilst confined to the upper halls of the temple, where his movements and his moods cannot be observed, rather than strutting and preening in front of his troops or casting divinations in the main hall. The general only adds value to the process of military engagement through his skilful manipulation of intelligence; if he allows information to flow both ways his utility is negated.

Though the essential function of the *Sunzian* general as a gatherer and analyser of information might seem to diminish him in comparison to the heroic aristocrat, the rhetorical structure of the text dispels such an interpretation. The concluding paragraph of the *Sunzi* leaves no doubt that it invests supreme significance in the role of the general:

> In ancient times, the Yin rose because Yi Zhi was with
> the Xia, the Zhou rose because Lü Ya was with the Yin.
> Thus only an enlightened ruler and an able general can
> use the supremely intelligent as spies, then they are
> certain of achieving great merit. This is the essence of
> [the use of] the military, that upon which the movement
> of the three armies waits.[23]

The transitional moments discussed here, the points at which one
royal dynasty gives way to another - the fall of the Xia dynasty and its
supplanting by Yin (commonly known as the Shang Dynasty) in the late
sixteenth century BCE and the Yin's ultimate demise at the hands of the
Zhou in the mid-eleventh century - were ascribed ultimate importance in
ancient China. In the ancient religion, the King and his family were the
only human beings empowered to maintain the offerings to Heaven; a
change of dynasty was therefore of momentous spiritual significance.
Among the Masters it was axiomatic that the moment of dynastic change
could reveal the secrets of good government. The last sovereign of a fallen
dynasty and the founding sovereign of its successor provided perfect
negative and positive models of kingship. These models were especially
urgent during the Warring States, when everyone was gripped by the sense
that another dynastic transition was both imminent and desperately
needed.

Against this context it is almost impossible to overestimate the
rhetorically provocative quality of the *Sunzi*'s assertion that dynastic
transition was achieved through the use of spies. This sense can be
illustrated by comparison of the *Sunzi* to the *Mencius*, a contemporary
Master's text. In this passage Mencius, a latter-day disciple of Confucius,
assesses the possibility that Qi, one of the most powerful of the Warring
States, could found a new dynasty:

> Even at the height of their power, the Xia, Yin and Zhou
> never exceeded a thousand li square in territory, yet Qi
> has the requisite territory. The sound of cocks crowing
> and dogs barking can be heard all the way to the four
> borders. Thus Qi has the requisite population. For Qi no
> further extension of its territory or increase of its
> population is necessary. The King of Qi can become a
> true King just by practicing benevolent government, and
> no one will be able to stop him.
>   Moreover, the appearance of a true King has
> never been longer overdue than today. It is easy to

provide food for the hungry and drink for the thirsty. Confucius said, "The influence of virtue spreads faster than an order transmitted through posting stations."

At the present time, if a state of ten thousand chariots were to practice benevolent government, the people would rejoice as if they had been released from hanging by the heels. Now is the time when one can, with half the effort, achieve twice as much as the ancients.[24]

Here Mencius candidly acknowledges stark historical differences between his own day and antiquity. The state of Qi is larger and more populous than the entire realm (that is, the entire civilized world) had been at the heights of the Three Dynasties of antiquity. One would imagine that this would make the task of founding a new dynasty harder than it had been in antiquity, as the six other feudal states with which Qi contended were equal to or greater than it in size. Mencius does not even entertain that notion, however, for he is confident that the key to establishing a dynasty remains unchanged from antiquity to the present day: virtue and benevolent government. The founders of the Yin and Zhou dynasties rose to kingship through the practice of morality. By practicing the same morality the ruler of Qi could bring the entire civilized world under his sway, and the civilized world had grown vastly since antiquity. This is why the King of Qi could "with half the effort, achieve twice as much as the ancients."

We can see how the *Sunzi*'s discussion of dynastic succession is an aggressive parting salvo against the perspective of a text like the *Mencius*. The *Sunzi* would agree with the *Mencius* that the key to success remains unchanged from antiquity to the present day. In the *Sunzi*, however, the priorities of the *Mencius* are inverted. The fact that the founders of the Yin and Zhou dynasty had less territory to conquer and less people with which to contend made their task easier according to the *Sunzi*'s criteria, yet despite that fact they had to rely on spies just as would a prospective unifier of the Warring States. Subterfuge, not virtue, was the key to the sages' triumphs.

In the choice of Yi Zhi and Lü Ya we see the author of the *Sunzi* making the same skilful rhetorical use of history exemplified by the attribution of the text to Sun Wu. Both men were celebrated in the Masters' literature as paragons of the wise minister, individuals of supreme talent and worth who had been drawn to the virtue of the sage founders.[25] In the *Sunzi* the exalted reputation of these men serves to underscore the difficulty of using them as spies. If success depends upon the effective manipulation of men with the talent of Yi Zhi or Lü Ya, command cannot be trusted to someone of mediocre competence. The *Sunzi*'s commander must be prepared and counted on to replicate the achievements of the sage kings of yore, and thus occupies a position of equal dignity and authority.

8.   Conclusion

We can thus see in the *Sunzi* a new paradigm that placed a premium on the rational and cognitive faculties of the commander and one that invents a new social role for the "general." Advocating expertise over superstition or good breeding is a laudable goal and much pursued in the modern world. Likewise the author is creating a divide (in some ways purposely artificial and polemically extreme) between his profession and his political overlords. It should be obvious that the author of the *Sunzi* is jealously protective of the institution of the professional general over which he claims ownership. His protectiveness is analogous to that displayed by professional commanders ever since, be they Clausewitz's contemporary Jomini, the archetypical professional, Helmuth Von Moltke or even the U. S. Army Command and General Staff School in the era following the First World War, as seen in the précis of this chapter. This is a common enough pattern for societies in the process of professionalising and routinising their institutions and those who staff them. The *Sunzi*, therefore, anticipates the kind of military-intellectual complex that all advanced societies manifest, and highlights many of the enduring tensions evident between the "professional" military and "amateur" statesmen.

Seeing too great a gulf between rulers and generals is but one example of the dangers inherent in intellectually slack readings of a text that was so sophisticated and rhetorically subtle in its original composition. The World War II German general, Günther Blumentritt, was famously credited with saying that giving a book as dense and complex as Clausewitz's *On War* to the military was akin to putting a razor blade in the hands of a child. We would not follow General Blumentritt in insulting the intelligence of contemporary military officers. Nonetheless, it is wise to remain vigilant when reading a subtle text that seems so simple and straight-forward. One of the greatest dangers of the conventional interpretation of the *Sunzi* as a storehouse of eternal wisdom or of the superficial reading of its aphorisms is that the rhetorical extremes to which the author goes to invent the "general" can be used to justify freeing the professional military from the bonds of political control. Such was not the author's intent. Rather he was seeking to define a realm within which expertise and intellectual ability were pre-eminent, but over which the interests of the state remained paramount.

Modern readers of the *Sunzi bingfa* will profit from keeping in mind the dialectical process of which the text itself was a part. As material realities changed on the battlefields of ancient China new problems and opportunities arose which could not be overcome or exploited in the absence of new roles and conceptions of command. In responding to this situation, the *Sunzi bingfa* changed the very dynamic of the world into which it was introduced, creating new problems and opportunities for later generations. A static reading of the *Sunzi bingfa* that treats it as an edifice of crystalline wisdom not only loses these dimensions of the text, but also overlooks the fact that the evolutionary process to which the text contributed continues today and implicates the modern reader as well.

Notes

[1] *Sunzi bingfa*, 3/3/3-6 (our translation). We employ the following version of the *Sunzi bingfa* as our baseline text: D.C. Lau, and Chen Fong Ching, eds., *A Concordance to the Militarists (Bingshu sizhong zhuzi suoyin [Sunzi, Yuliaozi, Wuzi, Simafa])* (Hong Kong: Commercial Press: ICS Series, 1992).

[2] *Sunzi*, 8/7/30.

[3] Antoine-Henri Jomini, *Précis de l'Art de Guerre*, Volume 1 (Paris: 1838), 135-136.

[4] Helmut von Moltke, *Militarische Werke*, Volume 2 (Berlin: E. S. Mittler, 1892), 291.

[5] *Principles of Strategy for an Independent Corps or Army in a Theater of Operations* (Fort Leavenworth: The Command and General Staff School Press, 1936), 19.

[6] *Sunzi*, 1/1/3.

[7] Qi occupied an area around current-day Shandong and Hebei Provinces. Wu comprised all, or parts, of today's Jiangsu, Anhui, Zhejiang and Jiangxi Provinces.

[8] Chu split modern Anhui with Wu and spread west and south over present day Henan, Hubei and Hunan.

[9] Yue included today's coastal provinces of Zhejiang and Fujian and inland territories in what is now Jiangxi Province.

[10] The use of the terms profession and professional to describe generalship in Ancient China are, to a certain extent, anachronistic. The author of the *Sunzi* is not inventing a "profession" in the modern sense; but the extent to which this book is forcefully delineating a unique realm of authority and expertise that is self-regulated by its members justifies the qualified use of

this terminology.

[11] *Sunzi*, 3/2/19-20.

[12] In an ideal system, the aristocracy of the Zhou was ranked in five positions below the king: *gong* (duke), *hou* (marquis), *bo* (earl), *zi* (viscount), and *nan* (baron).

[13] *Analects*, XV.1. D.C. Lau, tr., *The Analects* (London: Penguin, 1979), 132.

[14] Lau, 132.

[15] Yang Bojun, compiler and annotator, *Zuozhuan zhu* (Beijing: Zhonghua shuju), 198, Duke Huan Year 5.

[16] This is reflected in the graph for "army" (*jun*) itself, which is a picture of a chariot with a bracket over it to imply a "grouping of chariots."

[17] *Sunzi*, 2/2/14.

[18] *Sunzi*, 1/1/12.

[19] *Sunzi*, 1/1/20-21.

[20] *Sunzi*, 9/9/7-8.

[21] Michael J. Puett, *To Become a God: Cosmology, Sacrifice, and Self-Divinization in Early China* (Cambridge: Harvard University Press, 2002).

[22] *Sunzi*, 11/13/5-7.

[23] *Sunzi*, 13/14/26.

[24] D.C. Lau, tr., *Mencius* (London: Penguin, 1970), 76.

[25] Yi Zhi is known elsewhere as Yi Yin. He is mentioned in the *Mencius*, among many texts. Lü Ya is known elsewhere as Lü Shang or Grand Duke Wang of Lü. He is the putative author of the *Taigong liutao*, a later text on military strategy.

Notes on Contributors

Dorothea Flothow holds a Master's degree and studied English Literature and Modern History at the Universities of Tuebingen (Germany) and Reading (UK). While employed at the Sonderforschungsbereich "Kriegserfahrungen: Krieg und Gesellschaft in der Neuzeit" at the University of Tuebingen (a collaborative research centre sponsored by the German Research Foundation DFG), she completed her PhD dissertation on war imagery in children's literature. She currently holds a post-doctoral position at the University of Salzburg (Austria).

Albrecht M. Fritzsche received two Master's degrees from Albert-Ludwigs University in Freiburg, Germany, in Mathematics and Educational Studies. He has published several works on media, culture and science. He is currently a manager responsible for production planning systems in the IT department of a large American-German automotive company.

John C. Horn received his Master's degree in History from Sir Wilfrid Laurier University in 2004, a member of the Tri-University History Program (Laurier University and the Universities of Guelph and Waterloo), in Waterloo, Ontario, Canada. He is currently pursuing a PhD in First World War cultural studies.

Markku Jokisipilä who holds a doctorate in social sciences, works as a researcher and lecturer in the Department of Contemporary History at the University of Turku, Finland. His doctoral dissertation examined Finnish-German relations during the Second World War. His other publications have dealt with various aspects of political and military history, Finnish foreign policy, neo-conservatism, and sports and nationalism. The present article is based on his studies concerning online civil liberties and the effects of the Internet on politics.

Brigitte Le Juez is a senior lecturer in the School of Applied Language and Intercultural Studies at Dublin City University (Ireland). Her teaching, publication and research focus mainly on French literature and cinema. She is the Chair of the Master's program in Comparative Literature at DCU. Until 2000 she was also the President of the Association of French and Francophone Studies in Ireland, and is at present editor of *The Irish Journal of French Studies*.

Agnès Maillot is a lecturer in Intercultural Studies at Dublin City University, where she also teaches politics and history. Her main area of research is Ireland, and particularly Sinn Féin and the IRA.

Andrew Meyer received his PhD in East Asian Languages and Civilizations from Harvard University. For the past eight years he has taught courses on East Asian history and religion, and is currently assistant professor of history at Brooklyn College. In addition to his work on the intellectual history of the Tang Dynasty, he is collaborating with Andrew Wilson on a new translation of *Sun Tzu's Art of War*.

Mark L. Perry is an associate professor of sociology and cultural studies at Lebanese American University in Beirut. He pursues interdisciplinary studies in environmental issues, alternative culture, race relations, and the social implications of the telecommunications revolution.

Moni L. Riez is currently pursuing her PhD in Military and Strategic Studies at the University of Calgary. Her work centres on a comparison of the Hungarian and British home fronts during World War I. She also teaches high school social studies in Calgary.

Andrew R. Wilson holds a PhD in History and East Asian Languages from Harvard University. He currently serves as a professor of Strategy and Policy at the United States Naval War College in Newport, Rhode Island, where he lectures on military history and strategic theory. Recently he has been involved in editing a multi-volume history of the China War, 1937-1945, and is completing a new translation of *Sun Tzu's Art of War*.

Printed in the United States
by Baker & Taylor Publisher Services